ENERGY

a crisis — a dilemma — or just another problem?

Second Edition

ENERGY

a crisis — a dilemma — or just another problem?

Second Edition

Jesse S. Doolittle
Professor Emeritus
Department of Mechanical and Aerospace Engineering
North Carolina State University

Matrix Publishers, Inc.
11000 S.W. 11th Street
Beaverton, Oregon 97005

Library of Congress Cataloging in Publication Data

Doolittle, Jesse Seymour, 1903-
Energy : a crisis, a dilemma, or just another problem?

Bibliography: p.
Includes index.
1. Power resources. 2. Energy conservation. I. Title
TJ163.2.D65 1982 333.79 81-21393
ISBN 0-916460-33-9 AACR2

Production: Vicki L. Tobin
Illustrations: Scientific Illustrators
Printing: Pantagraph Printing
In-house Editor: Merl K. Miller
Series Editor: Francis J. Hale

Matrix Publishers, Inc.
11000 S.W. 11th St.
Beaverton, Oregon 97005

Contents

PART I
THE ENERGY SITUATION

1 **Introduction** 3

2 **Our Energy Demands, Past, Present and Future** 15

Introduction / Our Energy Demands and Energy Sources / Summary

3 **Electric Power Generation** 25

Importance of Electrical Energy / The Steam Power Plant

4 **Economics of Energy Production** 35

Fixed Costs / Operating Costs / Summary on Economics of Energy Production

PART II
OUR PRESENT ENERGY RESOURCES

5 **Introduction** 45

6 **Coal** 47

The Nature of Coal / Our Coal Resources / Mining Coal / Burning Coal / Peat / Summary on Coal

7 **Oil** 59

Oil and Its Products / The Importance of Oil / Our Oil Reserves / Summary on Oil

8 Natural Gas 65

Natural Gas Use / Natural Gas Resources / Summary on Natural Gas

9 Nuclear Power 71

Introduction / Nuclear Fission / Nuclear Fuel Preparation / Types of Reactors / Reprocessing and Waste Disposal / The Fast Breeder Reactor / Thorium and the Slow Breeder Reactor / Uranium Resources / Nuclear Reactor Safety / Summary on Nuclear Power

10 Hydroelectric Power 89

Hydropower / Present and Future Capacities / Pros and Cons of Hydroelectric Power / Pumped-Storage / Summary on Hydroelectric Power

PART III OUR ALTERNATIVE ENERGY RESOURCES

11 Introduction 101

12 Direct Solar Energy 103

The Solar Energy Resource / Direct Heating and Cooling / Passive Solar Heating / Direct Production of Electrical Energy / Indirect Production of Electrical Energy / Summary on Solar Energy

13 Wind Power 125

Introduction / Large Scale Development of Wind Power / The Future of Wind Power / Summary on Wind Power

14 Geothermal Energy 133

Introduction / Geothermal Steam / Geothermal Hot Water / Energy from Hot Rocks / Summary on Geothermal Energy

15 Marine Energy 147

Introduction / Tidal Power / Wind Waves / Ocean Currents / Thermal Gradients / Salinity Gradients / Summary on Energy of the Oceans

16 Shale Oil and Tar Sands Oil 159

Oil Shale / Shale Oil Production / Problems Associated with Shale Oil Production / Summary on Shale Oil / Tar Sands / Tar Sands Resources / Oil Production from Tar Sands / Summary on Tar Sands

17 Energy from Municipal Refuse 171

Municipal Refuse / Incineration / Pyrolysis / Summary of Municipal Refuse

18 Energy from Biomass 179

Biomass Resources / Wood / Vegetation / Gasohol / Other Biomass Resources / Summary on Biomass Energy

19 Nuclear Fusion 189

Introduction / The Fusion Process / Magnetic Confinement / Laser Fusion / Fusion Safety / Summary on Nuclear Fusion

PART IV ENERGY TRANSFORMATIONS

20 Coal Gasification and Liquefaction 201

Introduction / Gasification Concept / Power or Low Btu Gas / Industrial or Intermediate Btu Gas / Pipeline or High Btu Gas / Underground (In-Situ) Coal Gasification / Oil Gasification / Coal Liquefaction / Solvent Refining / The Status of Coal Gasification and Liquefaction / Summary on Coal Gasification and Liquefication

21 The Hydrogen Economy 213

Introduction / Hydrogen Production / Hydrogen Use for Energy Storage / Hydrogen for Energy Transportation / Using the Energy of Hydrogen / Summary on the Hydrogen Economy

22 New Conversion Methods 223

Introduction / Dual Cycles / Cogeneration / Magnetohydrodynamics (MHD)/ Fuel Cells / Summary on New Conversion Methods

PART V
CONSERVATION AND THE ENVIRONMENT

23 Energy Conservation 241

Introduction / Conservation in Electric Power Generation / Conservation in Industry / Conservation in Transportation / Conservation in Residential and Commercial Use / Summary on Energy Conservation

24 Energy and the Environment 259

Introduction / Energy Procurement / Energy Transformations / Energy Use / Acid Rain / Summary on Energy and the Environment

25 General Summary 273

An Overview / A Crisis? / A Dilemma? / Or Just Another Problem? / Conclusion

References, First Edition 279

Selected References, Second Edition 285

Appendices

A Energy Concepts and Energy Use 293

B The International System (SI) of Units 315

C Conversion Factors 317

D Properties 319

Symbols 321

Abbreviations 323

Index 325

Preface to First Edition

Since the Arab oil embargo in the fall of 1973, there have been very extensive discussions at all levels of our society as to the true nature of the energy situation in this country. Some feel that there is no real shortage of energy and that scare tactics have been used by the various energy companies to justify the large increases in costs of all forms of energy. Others have presented extensive information to show that the energy shortage is a very serious one and that it will become much worse unless steps are taken now to minimize it. The disparity between supply and demand could become so great as to cause a serious dislocation in our way of life.

Literature is voluminous with articles, pamphlets and books dealing with various aspects of the energy situation. Much of this literature is sound and informative. However, much of it deals with only one phase of the total problem and, in some cases, the information is biased. Although it is impossible to encompass the details of all phases of the energy situation in a single volume or even a series of volumes, an attempt will be made here to present the general aspects of the energy situation, present and future, in this country.

A study of the facts presented here should show that our energy situation is bad and getting worse. If we are to avoid a real catastrophe, it is important that government officials and business leaders make long-range energy decisions. It is essential that we, the citizens, understand the energy situation, that we have input before decisions are made, and that we support the decisions which are wise and seek to reverse the unwise ones.

It is not possible to write a book which is directed equally to all segments of our population. This book is designed as a text for students in community colleges and technical institutes, and non-engineering students in senior colleges and universities. Any faculty member with some background in science should be able to use this text to develop an interesting and profitable course.

In general, this book surveys our past and present uses of energy and attempts to show what our future demands will be. It then surveys the energy resources which now meet our demands. The so-called new energy sources are next examined to evaluate their potential for meeting our future energy needs. Consideration will be given to the possible effects conservation may have on reducing our energy demands. Finally, the interaction between energy production and the environment will be discussed. Illustrative examples will be given at the end of most chapters, followed by problems and questions for student use.

Although this country is moving towards the use of the metric or SI system of units, most people are not fully familiar with the new system. Hence, the units used through this text are the common English or engineering units. However, the SI system is discussed in some detail in the Appendix and conversion factors between the two systems are given.

Chapter 2 deals with energy concepts and energy transformations for the reader to understand more fully the potential as well as the limitations in the use of various forms of energy. For those concerned with only an overall view of the energy situation, this chapter can be omitted without seriously affecting the discussions in the remainder of the text.

Although this book is designed to be used as a text, it should prove valuable to all citizens who have only a very limited background in science.

Preface to Second Edition

Many developments have taken place in the energy field since the first edition was written. As a result, most of the tables in the first edition concerning energy supplies and predicted energy demands have been updated. The text material has been adjusted accordingly.

The author has attempted to incorporate the most important developments. In addition, material has been added which is of current interest, such as the use of gasohol, passive solar heating, cogeneration, solar ponds, and peat as fuel.

It is recognized that Chapter 2 of the first edition, Energy Concepts and Energy Use, has been a deterrent to some potential readers who lack an extensive background in Physics. Accordingly, it has been deleted in this edition. However, to accommodate those readers who wish to go rather deeply into the subject, the chapter has been re-written and placed in the Appendix. In order for the ordinary reader to understand some of the discussion which follows, certain fundamental terms and concepts have been added as Supplement C, Chapter 1.

The author continues to feel that the energy situation in this country is very serious, but too many of our citizens and some of our leaders form conclusions about energy without much knowledge of the true situation. The reader is invited to study all of the facts presented here and in other places if necessary. Then, and not until then, should conclusions be drawn.

The author is very grateful to two people for their invaluable help in preparing the manuscripts, Eleanor Bridgers for the first edition and Mary Hardee for the second edition.

PART I

THE ENERGY SITUATION

1

Introduction

Energy is the very lifeblood of our present civilization. Energy, obtained from nature, governs what we eat, what we wear, how we communicate, and how we transport our goods and ourselves. Energy controls the nature of our work, whether it be on the farm, in the factory, or in the office. But it has not always been this way.

Since the dawn of recorded history and up to recent times the changes in the way of life of mankind were exceedingly slow. To be sure, fire was discovered and, with wood as a fuel, metals were refined, tools and weapons of war were made. Over the centuries, there were a few remarkable engineering feats, such as the pyramids of Egypt and the Great Wall of China. However, these marvels were constructed by large segments of the population devoting their entire working lives to these projects.

In Europe, in the Middle Ages, great castles and cathedrals were built. Because of the contrast of these buildings with those built in previous centuries, it might appear that there had been a marked change in the way of life for the general populace. However, most of the populace were serfs whose lot in life was to build and defend the castles from enemies and to provide the necessities and luxuries for their lords and masters. Thus, even in the Middle Ages, the way of life of most of the population did not differ radically from that of many previous centuries.

The change in life styles came, almost imperceptibly at first, with the dawn of the industrial revolution in the latter part of the eighteenth and the beginning of the nineteenth centuries. This change became possible only when mankind learned to use energy other than muscular energy of man and beast. Steam power and water power were adopted to pump water from mines and, particularly, to drive machines, thus making it possible for vast changes to take place in the life styles of mankind.

As in other countries, changes in the way of life in this country were very slow at first. However, with the developments in the use of water power, coal, oil, and, later, natural gas, changes became much more rapid. In the last few decades abundant and low cost energy has made possible the high standard of living which we now enjoy. Generally, it was taken for granted that ever expanding supplies of energy would be available not only to maintain present life styles but to improve on them. The average citizen did not recognize that we were relying more and more on imported oil. Hence the Arab oil embargo in the fall of 1973 produced a dramatic shock on the country. Largely as a result of the embargo, long lines became common at gas stations. Other fuels, such as propane, heating oils, and natural gas were in short supply, causing a real energy crisis during the winter of 1973-74. This oil shortage was instrumental in initiating a serious recession.

With the removal of the oil embargo and recovery from the recession, much of the evidence of that crisis disappeared. The long lines at the gas pumps ended. Supplies of propane and heating oils again became seemingly ample to meet the demands. Major brownouts due to shortages of electrical energy did not occur. Unfortunately, because it appears that energy supplies are ample at present, a large segment of the population in this country has concluded that the energy crisis was largely contrived by the greed of the oil and power companies. This part of our population, encouraged by some politicians, has concluded that the proper steps to be taken would be to break up the oil companies and to nationalize the electric power companies. The assumption is made that then we could return, more or less, to the good old days of abundant and cheap energy.

An attempt will be made in this book to show that the truth of the matter is far different, a fact recognized by many individuals. Three typical views are quoted here:

1. Rocks and Runyon* state

> "An energy gap is opening up in America sufficiently wide to cause our total collapse in one decade."

2. Dr. Arthur M. Bueche,† Senior Vice President, Corporate Technology, General Electric Company, states

> ". . . unless we get our energy facts straight right now and tell the hard truth to our friends and neighbors and, especially, to the people we have elected to serve us, we could be in for a greater calamity by the year 2000 than wars, plagues, pestilence, or natural disasters have ever inflicted on this nation."

3. In a staff report entitled *The Geopolitics of Oil,* prepared for the Senate Committee on Energy and Natural Resources, November 1980, the following statement is made:

---

* Lawrence Rocks and Richard Runyon, *The Energy Crisis*. New York: Crown Publishing, Inc., 1972.

† Arthur M. Bueche, "The Hard Truth About Our Energy Future." Presented at the Western Conference of Public Service Commissioners, Seattle, 1979.

> "This gathering energy crisis deserves the highest priority in the council of governments. Few problems will be more difficult to solve. Moreover, many of the policies which we are currently pursuing to deal with the energy crisis are only making it worse."

Regardless of what should be done or what will be done to the oil and other energy companies, the fact remains that our energy supplies are running short. The days of abundant and cheap energy are gone forever. As our energy shortages increase, it is possible that drastic changes will occur in our way of life. (See Supplement A at the end of this chapter concerning the concept of "drastic change.") To prevent such drastic changes, painful decisions must be made not only by industry, but, particularly, by governmental agencies, including both Congress and the Executive branch of the government.

It is essential that the citizens of this country fully understand the energy situation in order for policy decisions to succeed in minimizing the effects of restricted supplies of energy. This means that the citizenry must be fully informed of the trends in our energy demands and our energy needs. Our present resources to meet current needs must be understood as well as the probability of these resources to meet our future needs. We must also examine the possibility of alternative sources meeting our future needs. Such alternative sources include

1. Solar energy
2. Wind power
3. Geothermal energy
4. Marine energy
    a. Tidal power
    b. Thermal gradients
    c. Waves and currents
    d. Salinity differences
5. Oil shale
6. Tar sands
7. Solid waste
8. Vegetation

It is the purpose of this book to examine the entire energy situation, presenting facts to help citizens formulate intelligent decisions as well as providing them with a full understanding of the problems involved in making these decisions.

The following facts are presented to help the reader realize that we face a severe and ever growing energy problem.

1. We as a nation, having less than 6 percent of the world's population, consume about one-third of the energy used by all of the nations of the world.
2. In one year we use the energy which it took nature a long, long time, perhaps a million years, to create.

3. We are now importing over 40 percent of the oil we are using. Our oil imports are approximately three times as large as in 1960. It is costing the country at least 90 billion dollars a year for this oil, thus making a significant impact on our economy. There is reason to believe that the high cost of imported oil contributed to the recession of 1974-1975.
4. For many years oil could be readily obtained by drilling, at the most, to a few thousand feet. At present, most oil and gas wells are several thousand feet deep and, in several cases, wells are now several miles deep. To meet demands we have been forced to seek oil and natural gas off shore. Today approximately 13 percent of our domestic oil is obtained off shore, largely in the Gulf of Mexico.
5. Approximately 72 percent of our energy demands are being met by the use of oil and natural gas. In spite of very deep well drilling on land and increasing drillings off shore, domestic production of oil and natural gas leveled off and now has started to decrease.
6. Our demands for energy are ever increasing. We used approximately 2½ times as much energy in 1975 as we did in 1950.
7. Although our supplies of coal and nuclear fuel are large, many problems are involved in their use. Furthermore, the costs of these fuels have been rising rapidly.
8. As will be discussed in detail, we have not been able to produce significant amounts of energy from other sources. If ultimately we can do so, this energy will be expensive.

An examination of these facts shows that the days of abundant and cheap energy are passed. Since this country is so dependent on large supplies of energy, decisions must be made to prevent what most of our citizens would feel to be a serious deterioration in our standard of living. Several possible courses of action are open. These will be discussed briefly. As we examine the total energy situation, it should be apparent that no one course of action will, in itself, be wholly satisfactory.

1. We can continue to drift without any real policy being formulated. Without a policy we will import ever increasing amounts of oil. However, other nations, particularly the energy-poor nations, will compete with us for the available oil supplies, thus driving up the costs. Assuming that the oil is available for us to import, it has been predicted that imported oil will cost us at least 100 billion dollars a year in 1985. This means a cost of 425 dollars a year for every man, woman, and child in this country. The payment of such a vast sum of money will have a drastic effect on our economy.

2. We can greatly reduce our energy demands by government edict. We can ration not only gasoline but also fuel oils and electricity. We can outlaw labor-saving, energy-hungry machines. We can shift a large segment of the population back to the farms to replace energy-intensive machines. We can eliminate energy consuming devices in our homes. We can increase the work week to 60

or more hours as man replaces machines. Thus we can return to the "good old days" where man lived by the sweat of his brow.

3. We can, by either suitable rewards to the oil and gas companies or by government action, greatly increase the production of oil and natural gas. But such action will only temporarily relieve our energy problem. The more oil and natural gas we use now, the less oil and gas will be available for future use. Thus, we would deprive not only ourselves but future generations of these fuels. Our tire industry, our plastic industry, our fertilizer industry, and other chemical industries depend on feedstocks derived from oil and natural gas. The burning of these fuels now will rob future generations of these very valuable feedstocks.

4. We can greatly expand the use of coal and nuclear fission energy. We have large reserves of coal, but expansion in our use of it will create major problems in financing the expansion, in mining the coal, in transporting it, and burning it. Many scientists feel that nuclear fission energy, particularly when the fast breeder reactor is developed, will make a very major contribution to relieving our energy shortage. But many earnest citizens, including some nuclear scientists and engineers, feel that the problems of nuclear safety are so great that further development of nuclear fission should be halted until these problems can be resolved. There is some feeling that all use of nuclear fission should be abandoned completely.

5. We can develop alternative sources, particularly solar, geothermal, wind, shale oil, and nuclear fusion. There is wide disagreement as to the potentials of the new alternative sources. It is fairly well agreed that nuclear fusion will not become a practical reality in this century. However, because of the tremendous possibilities it offers, research continues. Many people feel that with proper research and development efforts, solar, geothermal, shale oil, and wind power can greatly contribute towards meeting our energy needs. Others feel that none of these four energy sources will ever make significant additions to our supply of energy. In any case it does not appear that we can expect significant amounts of energy from these sources in the next decade or two.

6. We can reduce our energy demands significantly but not so drastically that there will be a major change inour standard of living. At the same time we can take steps to produce more oil and natural gas and to expand our production and use of coal and nuclear energy. Thus, we should be able to satisfy our energy needs for the near term (ten to fifteen years). We could also expend much effort to develop the alternative sources for the more distant future.

Assuming that we do not curtail our demands for energy drastically, two very major questions require answers, independent of how we increase our energy supplies.

1. How can we protect our environment as we extract energy from nature and put it into the desired form? In most cases, the protection of the environ-

ment will be very expensive. How far shall we go in a compromise between the degree of protection of the environment and the cost of this protection?

2. How can we secure the vast sums of money required to develop additional energy sources and at the same time protect the environment? Should we permit the energy companies to earn relatively high profits* so they will have the resources to produce additional energy supplies and also attract investors to supply additional capital? On the other hand, have the various governmental agencies demonstrated their capabilities to produce energy and to protect the environment better and at lower costs than private industry?

In general, enormous sums of money and other resources will be required to search out and to develop additional energy sources and also to protect the environment. The public must pay the bill either through the energy it buys or through increased taxes if the government is to be the producer of the additional energy supplies.

Unless we are to drift into a deterioration in our way of life, important decisions must be made as to the optimum methods of securing needed energy supplies, how to pay for this energy, and how to protect the environment. Intelligent decisions cannot be made without a full understanding of all aspects of the energy situation.

In discussing various energy sources, consideration must be given to:

1. The potential of each source to help satisfy our energy needs.
2. The cost factors involved in the development of these sources.
3. The adverse environmental effects caused in the development of these sources.
4. Methods and costs of minimizing these adverse environmental effects.

SUPPLEMENT A

POSSIBLE DRASTIC DETERIORATION IN OUR WAY OF LIFE

If we were to drastically reduce our energy use, say to one-half of what we are now using, of necessity we would change our life styles back to those of several decades ago. In factories much of the automation would be missing, with many tasks being performed manually. On farms, the large energy-intensive machinery would be missing. In its place would be simple machines, together with horsepower, mule-power, and manpower. In our homes many of the labor saving devices would be missing. The housewife would find it necessary to devote much more of her time to running the household. Particularly with larger families, it would be more difficult for the wife to be employed outside the home.

---

* See Supplement B for a more detailed discussion of the need for money for energy development and company profits.

Instead of two or three cars per family, we would have only one car or none at all. Our ability to come and go at will would be greatly restricted. We would be tied to our jobs to a much larger extent. Our vacation travels would be much more limited.

It may be true that even under those conditions, our standard of living would exceed that of most of the world today. Some people could adjust rather readily to a change in their life style and perhaps might enjoy a more simple life. Others, particularly those who have moved up from the near poverty level or below, would find it very difficult and most undesirable to return to their previous status. It would be very difficult for the factory worker and the farm worker to return to back-breaking manual labor once they had enjoyed the advantage of machines to perform the harder jobs.

Many of us would find it very difficult to adjust to a drastic reduction in energy available to use in our homes. For instance, consider air conditioning. Thirty years ago, very few homes were air-conditioned. Although we were uncomfortable in the summer, we accepted this situation as a normal way of life. Today, if we were deprived of air conditioning in our homes in the summer, we would find the situation unbearable.

Thus there is much reason to believe that most people would find it very difficult to give up their present way of life to which they have become accustomed. Unless the reduction in energy supply were to be made gradually over a period of many years, there would be a great outcry about a large reduction in the energy available to us.

SUPPLEMENT B

FINANCING OF DEVELOPMENT OF ADDITIONAL ENERGY SUPPLIES

Unless we curtail our energy demands drastically, it will be necessary to expend vast sums of money to expand our conventional energy sources. Much money will be required to open new coal fields, to protect the environment during mining, and to expand our coal transportation systems. Very expensive equipment will be required to protect the environment where coal is burned. Billions upon billions of dollars will be required to construct additional conventional power plants. Enormous sums of money will be required to drill for new supplies of oil and natural gas, particularly in the deep oceans.

As will be discussed later on, there are many proposed alternative energy sources such as solar energy, wind energy, geothermal energy, biomass energy, nuclear fusion, marine energy, oil shale and tar sands sources. Before any of these alternative energy sources can make a major contribution to our energy supply, large sums of money must be spent to establish the economic feasibility of developing these energy sources. Once the economic feasibility is established, then billions of dollars will be required to establish full size commercial plants.

It must be recognized that unless these sums of money become available, the alternative sources will not supply significant amounts of energy no matter how much we demand it. Unless we, as energy consumers, will pay the bill, either directly or indirectly, these energy sources can not be developed.

Where can this money come from directly? If we feel that the federal government has demonstrated its ability to carry out projects like these in the most efficient manner, then the federal government should do the job. We will then pay for the new energy we receive directly or through taxes. If, on the other hand, we feel that private industry is better qualified to develop these energy sources, we must ask how the money is to be obtained. We may think of many companies, particularly the larger oil companies, as being rich, all powerful, and with almost unlimited funds. In spite of their large resources, in themselves these large energy companies do not possess the vast sums of money required. Funds can be obtained by plowing back some of the profits or by raising capital from investors. Most new or additional sources of energy involve high risk. Hence, it is difficult to attract investors, particularly if the company does not invest much of its own money—money obtained from profits.

The subject of the reasonableness of profits is generally misunderstood, particularly as the news media frequently report changes in profits rather than total profits. This leads to erroneous conclusions. Consider for example, that Company A has a poor year and earns only 1 percent of its invested capital. The next year it earns 3 percent. The news media probably would report that Company A had a banner year as its profits increased 200 percent. On the other hand, assume Company B had a good year and made a profit of 20 percent. The second year its profit is 21 percent. The profit increased only 5 percent. Is that bad?

It is essential that we look primarily at the total profit and not the year-by-year change. Two questions should be asked:

1. Is the total profit reasonable? Is it in line with profits from other businesses?
2. What is done with the profit? Do the stockholders receive excessive amounts or is a large percentage of the profits used to search for additional sources of energy to satisfy our needs?

Unless we are willing to pay increased taxes to the federal government to develop additional energy sources or unless we allow the energy companies to make reasonable profits which they will use for energy development purposes, the energy which we want in the future will not be available to us.

SUPPLEMENT C

DEFINITION OF TERMS

A full understanding of the nature of energy and energy related quantities is necessary to fully comprehend the energy situation. Recognizing that some readers may lack the minimum background in Physics to study this material, it has been placed in the Appendix. However, it is desirable to have some concept of certain fundamental terms if the discussion of energy is to be rational. Refer to the Appendix for more details.

In the narrow sense, *energy* is defined as the capacity to do work. In the broader sense, energy is associated with any and all changes in the physical or chemical aspects of matter. In the English system of units, energy may be expressed in *British Thermal Units* (Btu) or the *foot pound* (ft. lb.). As a close approximation the Btu can be defined as the amount of heat required to increase the temperature of one pound of water one degree Fahrenheit (°F) when the water is at room temperature. When a force of one pound acts on an object causing it to move through a distance of one foot, the work done is one foot pound (ft. lb.). As a close approximation, 778 ft. lb. = 1 Btu.

A term encountered very frequently in dealing with energy is *power*. Power is defined as the time rate of doing work, i.e., the amount of work done in a specified time. A *horsepower* (hp) is defined as 33,000 ft. lb. of work done in a period of one minute. Since there are 60 seconds in a minute, one horsepower represents 550 ft. lb. of work done in one second. Likewise, when 1,980,000 ft. lb. of work are done in an hour, the power involved is one horsepower. Since 1 Btu equals approximately 778 ft. lb., a horsepower hour equals 2,545 Btu.

The product of volts and amperes equals *watts* for direct current electricity. Since the watt is the time rate of flow of electrical energy, it is a unit of power. For larger quantities of electrical power the unit used is the kilowatt (kW), which is 1,000 watts. One horsepower is equivalent to 746 watts. Then a kilowatt is approximately equivalent to 3,413 Btu per hour. When a time is associated with the kilowatt, it becomes a unit of energy. Thus a kilowatt hour (kWh) is approximately equivalent to 3,413 Btu.

Energy exists in many forms. When energy transformations do take place, there is an exact relationship between the various energies involved. When work is used to overcome friction, the heating effect is exactly equal to the work used. Likewise when heat is used to develop work, there is a precise relationship between the heat used and the work produced. In a nuclear fission process, mass is transformed into energy. There is an exact relationship between the energy produced and the mass transformed. These statements of the equality of energies in all energy transformation processes are known as the *first law of thermodynamics*.

It is very easy to cause some energy transformations to take place. When electric currents flow in conductors, a heating effect is produced. When work is transmitted through mechanical devices, some of it is lost because of friction. This produces a heating effect. On the other hand, it is difficult to transform heat into work. Most of our electrical energy is produced by transforming heat into work and then, by use of electrical generators, into electrical energy. The transformation of heat into work is accomplished in so-called heat engines. The steam turbine power plant is a common example of a heat engine plant. Here heat is added to produce steam, with the steam developing work in a steam turbine. After developing the work, the steam rejects heat at a low temperature as it is condensed in a condenser.

The term *thermal efficiency* is used to measure the amount of heat supplied which is turned into work. Thus, thermal efficiency,

$$\eta_t = \frac{\text{Work}}{\text{Heat supplied}} \tag{1-1}$$

The amount of heat which can be turned into work depends on the temperatures at which it is supplied and rejected. The ideal (the maximum possible) thermal efficiency is given as

$$\text{Ideal } \eta_t = \frac{T_h - T_l{}^*}{T_h} \tag{1-2}$$

T_h=The absolute temperature† at which heat is supplied.
T_l=The absolute temperature at which heat is rejected.

Because actual heat engines are far from being perfect, the actual thermal efficiency is much less than the theoretical. The limitation on the ability to transform heat into work is one aspect of the second law of thermodynamics. There is no way of mathematically proving the limitation but all of experience verifies it. In fact no one has been able to devise, even theoretically, a heat engine having a higher thermal efficiency than that just stated.

Since so much of our energy is derived from fossil fuels (coal, oil, and natural gas), the term *heating value* will be used extensively. Assume that an air-fuel mixture is supplied to a combustion device at a specified temperature. The fuel is burned completely and the products of combustion are cooled down before they leave the combustion device to the temperature of the original air-fuel mixture. If the fuel contains hydrogen it is assumed that the water vapor formed by burning the hydrogen is completely condensed before it leaves the combustion device at the entering temperature. The total amount of heat removed from the combustion device is known as the higher heating value at constant pressure of the fuel. Unless otherwise stated, the term heating value as used throughout the text will be this value. (There are some other heating values which are useful under some specified conditions.)

In addition to energy terms, two other terms will be used extensively. These are *temperature* and *pressure*. The temperature most commonly used in this country is the *Fahrenheit temperature* (°F). Under certain conditions it is necessary to express temperatures above the lowest possible temperature. These temperatures are known as absolute temperatures and may be expressed as degrees *Rankine* (°R). The absolute temperature may be determined from the Fahrenheit temperature as follows:

$$°R = °F + 460‡ \tag{1-3}$$

---

* This is known also as the Carnot cycle thermal efficiency.

† Absolute temperatures will be discussed shortly.

‡ The precise value is 459.69.

The so-called *Celsius* (sometimes called Centigrade) scale is used extensively. These temperatures are designated as °C. The absolute temperature using this scale is known as degrees *Kelvin* (°K). This absolute temperature may be found as follows:

$$°K = °C + 273^{*} \tag{1-4}$$

The relationship between the Fahrenheit and Celsius scales is shown in Figure 1-1. In equation form,

$$°C = \tfrac{5}{9}(°F - 32) \tag{1-5}$$

$$°F = \tfrac{9}{5}°C + 32 \tag{1-5a}$$

The relationship between Rankine and Kelvin temperature is:

$$°K = \tfrac{5}{9}°R \tag{1-6}$$

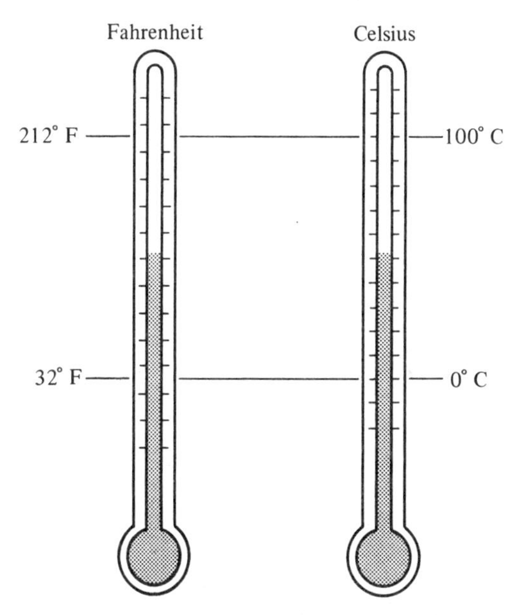

Figure 1-1. Fahrenheit and Celsius thermometers

Pressure may be expressed in *pounds per square inch* (psi). Ordinary pressure gauges register pressures above or below atmospheric. These pressures are designated as pounds per square inch gauge (psig). Under some conditions, *absolute pressures* (pressures above that of a complete vacuum) are required. This pressure may be obtained by adding the atmospheric (sometimes called the barometric) pressure to the gauge pressure. The absolute pressure is designated as psia. Unless precision is desired or for operating at high elevations, the atmospheric pressure may be taken as 14.7† psia.

* The precise value is 273.15.

† The precise value is 14.696.

2

Our Energy Demands, Past, Present and Future

2.1 INTRODUCTION

In Chapter 1 it was stated that we in this country, with less than six percent of the world population, are now using about one-third of the energy used by all nations of the world. As shown in Table 2-1, in 1970 we used approximately 50 percent more energy than in 1960. Part of the increase is due to an increase of population which grew by about 14 percent. The major cause of our national increase in the use of energy was the increase of 31 percent in the per capita use of energy.

Much of the large increase in energy use has been caused by the shifting of work loads from man to machines. This has been particularly true on the farms where mechanization has largely displaced hard labor. As machines replaced manpower, there has been a large movement of population from the farms to urban areas. Since the small farmer has been unable to afford large expensive machines, many small farms have disappeared.

There has been, also, a large increase in the amount of energy used in our factories as automation displaces hand labor. This growth in automation has increased the output of manufactured goods but it has decreased the need for factory workers, thus contributing to unemployment. It has been suggested that we decrease the amount of automation and that we shut down many of our large, energy-hungry machines, thus creating many more jobs. Although this would help minimize the unemployment problem, factory work again would become real drudgery.

Another reason for the large increase in the use of energy in this country has been that the American people have come to feel that energy is almost inexhaustible and is cheap or ought to be. As a result, we have been very wasteful. We demanded large, gas guzzling cars. We insisted on our right to use our cars whenever and wherever we desired. To minimized building costs, we did not install sufficient insulation and other energy-conserving devices in our homes, old and new. We paid little attention to efficient use of energy in our homes, in our cars, in our factories, and on our farms.

It is true, primarily because of high energy costs, some real efforts, particularly by large industries, are being made now to conserve energy. These conservation efforts, together with the reduced industrial activities during the 1974-75 recession, have slowed our energy demand annual growth to 1 to 2 percent instead of the previous rate of 4 to 5 percent. With greatly increasing costs of fuels and electricity, the general public now has been forced into conservation of energy. Storm windows and doors have been added to a large number of homes and much insulation has also been installed. By government edict as well as by popular demands, automobiles have become smaller and much lighter resulting in a significant reduction in the demand for gasoline. In the future, as older gas guzzling cars are phased out, the amount of gasoline used per car will decrease. On the other hand, with an increase in population, there will be a sizeable increase in the number of cars on the road.

It is essential that we examine our growth patterns in order to plan to meet our anticipated demands for the future. Unless this is done, it is possible that the chaotic conditions predicted by some will occur.

In making such an examination, it is desirable to again emphasize the three factors responsible for the very large increase in our demands for energy in the past and which will control our energy demands in the future:

1. A significant increase in our population.
2. A large increase in the use of labor saving devices.
3. Our very wasteful use of energy.

2.2 OUR ENERGY DEMANDS AND ENERGY SOURCES

Before presenting data concerning our energy demands and energy sources, several factors should be discussed. First it takes time to collect data, to assemble it and publish it. Hence data for a given year may not appear in print for a year or two.

It is very difficult to determine precisely the number of forms of energy which have been used and the distribution of this energy. The data presented here appears to be the most reliable and is taken primarily from government sources.

In order to plan for the future, it is necessary to have some idea of our future energy demands. However, it is very difficult to predict these demands. They will depend on the economic situation, the availability and costs of conventional forms of energy, and the progress made in development of alternative energy sources. Certainly we cannot expect the same growth in the use of energy which occurred in the last twenty-five years, but we should expect an appreciable growth in demand for the following reasons:

1. There will be a population growth.
2. People presently with low incomes and low energy use will press for a better lot in life and can be expected to demand more energy.

3. As our industry becomes more automated and as large farms use larger farm implements, the demand for energy will increase.

Two terms are used commonly in reporting energy uses. One of these is *barrels of oil equivalent*. Unfortunately there is not an agreement as to the energy value of a barrel of oil. In some cases, the oil is assumed to be crude oil having a heating value of 5,800,000 Btu per barrel. In other cases, a refined oil is considered, with a heating value of 5,500,000 Btu per barrel. An alternate method of evaluating the amount of energy is to express it in *Quads*. A Quad (short for quadrillion) is 10^{15} Btu. It is roughly equivalent to 177,000,000 barrels of oil.

In the tables and figures which follow, data about past use of energy will be presented, based largely on the *Annual Report to Congress, Volume 2*, by the Energy Information Administration, U.S. Department of Energy, 1979. The predicted future use of energy presented here is based largely on "Synopsis of Energy, Facts and Projections," from the *Annual Report to Congress*, by the Energy Information Administration, U.S. Department of Energy, 1978. This report makes three sets of predictions for the future. The medium projection figures are presented in the tables of this text.

Presently, as has been true for some time in the past, our energy demands are being met largely by the use of fossil fuels, a non-renewable source. A more recent nuclear fuel, uranium, is starting to supply some of our energy. It is anticipated that by the year 1995, uranium will supply about 11 percent of our total energy. Hydropower, a renewable energy source, supplies this country with a small amount of energy. However, in certain regions of the country, such as the Columbia river basin, the Colorado river basin, and the Tennessee Valley, hydropower supplies a significant portion of the energy demands.

Table 2-1 shows the U.S. energy supply for the years 1960-1979. It also shows the predicted energy supply up to the year 1995. This data is plotted in Figure 2-1. To have a valid basis of comparison, the hydro energy listed in Table 2-1 is the equivalent amount of fossil fuel energy required to produce the electrical output of the hydro plants. This means, for instance, if in 1979 we had no hydropower, we would have to use 3.16 Quads of fossil fuel to replace it. This is approximately 4 percent of the total energy used in that year.

Several pertinent facts can be obtained from Table 2-1:

1. Up to the present time, most of our energy demands have been met by fossil fuels, a non-renewable resource. In 1979, 92 percent of the energy which we used came from fossil fuels.

2. The domestic production of natural gas has reached a peak and will decline in the near future. We are importing an increasing amount of natural gas. Much of this imported gas will be LNG (liquified natural gas). LNG is produced by cooling natural gas down to at least 250°F below 0°F and subjecting it to high pressure. It becomes a liquid and then is transported in especially designed ships.

3. Domestic production of oil reached a peak around 1974. Recently production has increased somewhat; it is predicted that this small increase will

Table 2-1. Energy Supply in U.S. (Quads per year)

		1960	1970	1975	1979	1985	1990	1995
Oil	Domestic	16.39	22.91	20.10	20.40	21.79	23.07	23.96
	Import	4.00	7.47	12.95	17.64	17.42	16.96	16.36
	Total	20.39	30.38	33.05	38.04	39.21	40.03	40.32
	Used	19.92	29.52	32.73	37.02			
Natural gas	Domestic	12.60	21.67	19.64	19.19	18.19	17.40	17.08
	Import	0.16	0.85	0.98	1.27	1.88	2.04	1.57
	Total	12.76	22.52	20.62	20.46	20.07	19.44	18.65
	Used	12.39	21.79	19.95	19.86			
Coal	Domestic	11.12	15.05	15.19	17.41	22.69	31.22	41.70
	Export	1.02	1.94	1.79	1.78	1.90	2.08	2.28
	Total	10.10	13.11	13.40	15.63	20.79	29.14	39.42
	Used	10.12	12.66	12.82	15.08			
Nuclear		0.01	0.24	1.90	2.75	6.57	9.43	12.66
Hydro	Produced	1.60	2.63	3.15	2.96			
	Used	1.65	2.65	3.22	3.16			
Others, including hydro						3.24	3.49	3.97
Total Used		44.09	66.83	70.70	78.02			
Total Supply						89.88	101.53	115.02

Note: The total net supply may not agree with that used because of losses and changes in the amounts in storage. Some electrical energy is imported from Canada.

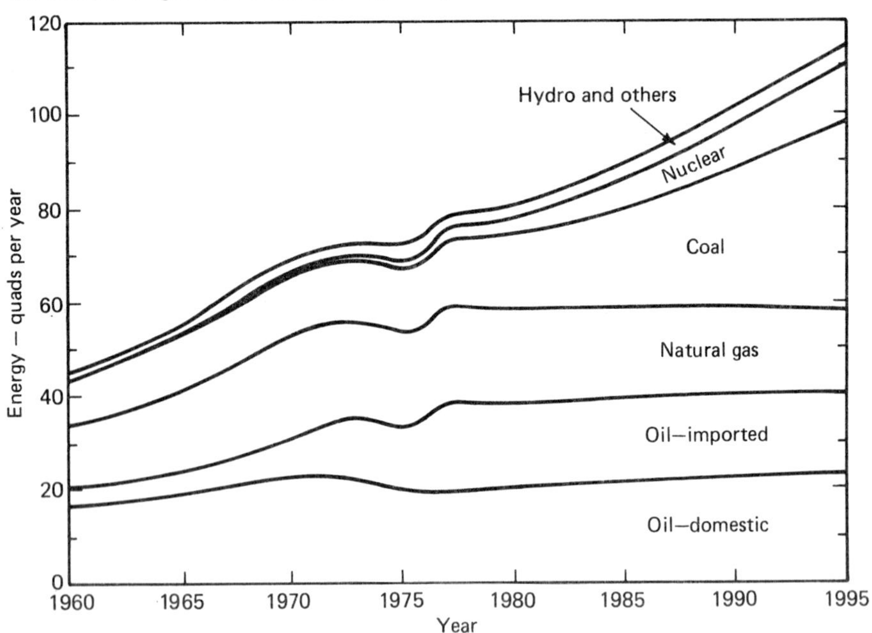

Figure 2-1. U.S. energy supply

continue for some time. This increase is due largely to Alaskan oil, though large increases in prices have made it feasible to use secondary and tertiary methods to obtain more oil from existing wells. According to Table 2-1, the amount of imported oil will gradually decrease, due to the depletion of foreign fields, increases in costs, and to demands for oil by other countries.

4. Although not shown in Table 2-1, in the 1950's there was a decrease in the domestic use of coal. Now, with the growing shortages of oil and natural gas, and with greatly increasing energy demands, coal is being used more extensively. It is anticipated that the use of coal will continue to increase.

5. In spite of the many problems associated with its use, nuclear energy is satisfying an increasing percentage of our needs. Table 2-1 shows that nuclear energy will provide 11 percent of our energy supply in 1995, compared to less than 4 percent at present.

6. Years ago, hydropower supplied a significant percentage of energy to meet demands. In certain regions of the country, such as the Columbia river basin, the Colorado river basin, and the Tennessee Valley, hydro now satisfies a large portion of the energy needs. For the country as a whole, however, hydropower, even when combined with newer forms of energy, will supply a smaller percentage of our energy requirements, in spite of a slight increase in hydropower production. The demands for energy will increase much faster than production.

Table 2-2 shows the distribution of the energy uses between various segments of society. This information is plotted in Figure 2-2. Note that the energy listed under "Industrial" includes the energy of fossil fuels used to produce such items as plastics. Table 2-2 shows that conversion and line losses (electrical) increase much faster than do the other uses due to the great increase in the use of electrical energy (see Table 2-4). It is predicted that the future conversion and line losses will be a somewhat smaller percentage of the energy supplied for electric generation than at present.

Table 2-3 shows the source of energy for the electric utilities. This information is plotted in Figure 2-3. Note from Table 2-3 that the use of natural gas for electric generation reached a peak about 1975 and is predicted to drop drastically in the future. It is anticipated that the use of oil for electric generation will increase for some time but will decrease significantly by 1995. It is

Table 2-2. Energy Consumption Distribution (Quads per year)

	1960	1970	1975	1979	1985	1990	1995
Residential and Commercial	11.78	17.64	17.10	18.89	18.32	19.30	20.33
Industrial	15.83	21.49	20.21	21.80	28.23	33.18	39.44
Transportation	10.58	16.21	18.38	19.98	21.00	21.89	23.32
Conversion and Line Losses (Electrical)	5.90	11.54	14.52	17.32	22.33	27.17	31.95
Total	44.09	66.83	70.71	77.99	89.88	101.54	115.04

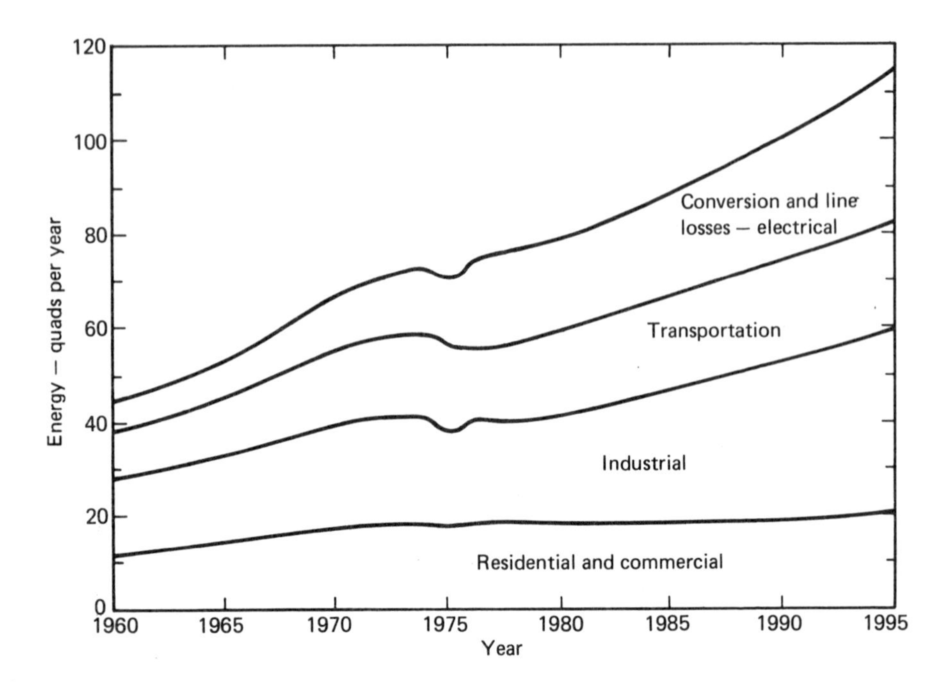

Figure 2-2. Energy consumption distribution

Table 2-3. Energy Consumption by Electric Utilities (Quads per year)

	1960	1970	1975	1979	1985	1990	1995
Coal	4.24	7.24	8.79	11.23	15.44	22.05	28.44
Natural Gas	1.79	4.05	3.24	3.61	2.05	0.51	0.04
Oil	0.57	2.13	3.23	3.53	4.47	3.17	0.58
Hydro (Fossil Fuel Equivalent)	1.57	2.60	3.12	2.92	3.05	3.15	3.24
Nuclear	0.01	0.24	1.90	2.75	6.57	9.43	12.66
Geothermal, Wood Waste, etc.		0.02	0.07	0.09	0.19	0.34	0.74
Total	8.18	16.28	20.35	24.13	31.77	38.65	45.70
Electric Sales	2.34	4.75	5.90	7.03	9.41	11.46	13.71

predicted that the amount of power generated in hydro plants will decrease from 19 percent of the total in 1960 to about 7 percent in 1995. On the other hand, it is predicted that the amount of power generated in nuclear plants will increase from 11 percent in 1979 to over 25 percent in 1995. In 1995 we will depend on coal and nuclear energy for over 85 percent of our electrical energy. Although the use of geothermal, wood, and waste energy sources will increase, their total supply for electrical energy is predicted to be less than 2 percent of the total for electrical energy generation in 1995.

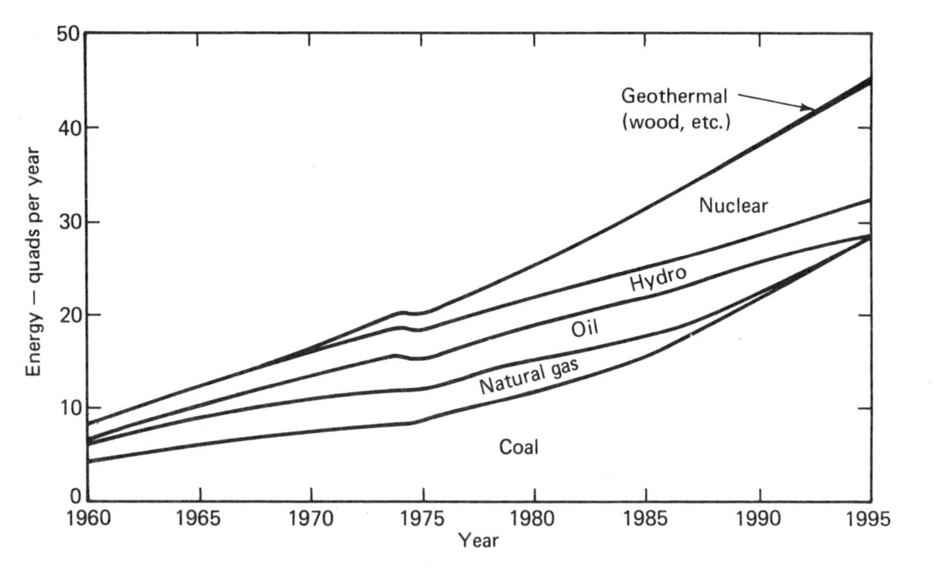

Figure 2-3. Energy consumption by electric utilities

Table 2-4 and Figure 2-4 show the end use of electrical energy. The rapid increase is expected to continue. As industry becomes more and more automated, its demand for electrical energy will increase quite rapidly.

Although not shown in the tables, the four figures in this chapter show the effects on energy consumption of the Arab oil embargo in the fall of 1973 and the ensuing recession.

Table 2-4. End Use of Electrical Energy (Quads per year)

	1960	1970	1975	1979	1985	1990	1995
Residential	0.69	1.59	2.00	2.32	2.83	3.34	3.82
Commercial	0.41	1.05	1.37	1.61	2.61	3.09	3.50
Industrial	1.12	1.95	2.30	2.85	3.96	5.02	6.38
Others	0.11	0.16	0.23	0.25	0.01	0.01	0.01
Total	2.33	4.75	5.90	7.03	9.41	11.46	13.71

2.3 SUMMARY

1. Until the recession of 1974, our demands for energy increased very rapidly. In 1970 we used over 50 percent more energy than in 1960. Our future demands for energy will continue to increase but not as rapidly as in the past.

2. In 1979 we depended on oil and natural gas to supply 72 percent of our energy needs. This percentage is predicted to decrease to around 50 percent in 1995.

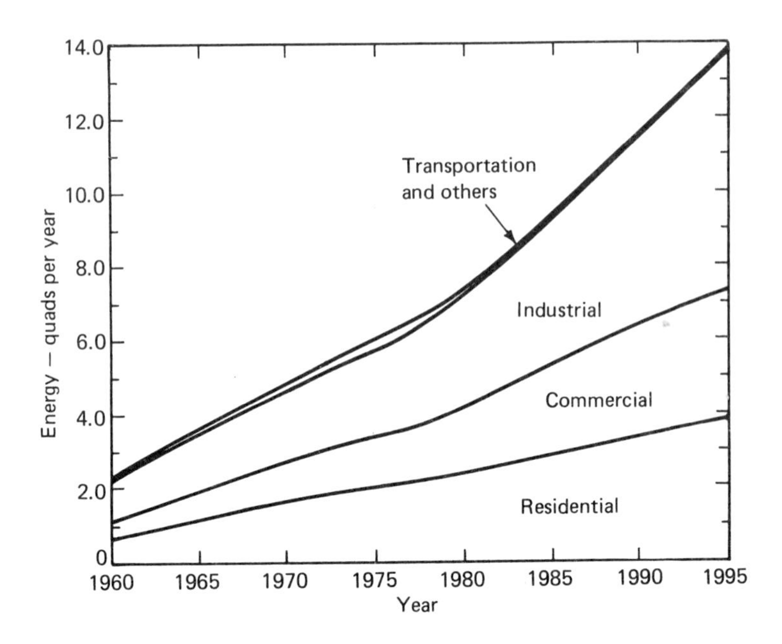

Figure 2-4. End use of electrical energy

3. We are importing about 45 percent of the oil which we use. The total amount of imported oil may decrease somewhat but its greatly increased cost will make an impact on our economy.

4. In spite of pressures from environmental groups we must depend on nuclear power and coal to supply a large portion of our energy for at least two decades. The alternative is to drastically reduce our use of energy.

SUPPLEMENT

PROBLEMS

2.1. Assuming that the average heating value of coal is 12,500 Btu per lb. and using Table 2-1, determine the tons of coal used in this country in 1960 and in 1979.

2.2. Using Table 2-1, determine the percentage of our energy use in 1979 derived from (a) oil, (b) natural gas, (c) coal, (d) nuclear, and (e) hydro and others.

2.3. Same as Problem 2-2 except for the estimated energy supply in 1995.

2.4. a. Using Table 2-2, calculate the ratio of the total predicted energy consumption in 1995 to that in 1979.
  b. Using Table 2-3, calculate the ratio of the predicted electrical energy sales in 1995 to that in 1979.
  c. Comment on the answers to parts a and b.

2.5. Defining the overall efficiency in electrical power production and distribution as the ratio of sales to energy consumption by the electric utilities, determine this efficiency for the years 1960 and 1995, using Table 2-3.

2.6. Determine the per capita use of energy Btu per year in 1960 and the projected for 1995. Use Table 2-1. (The U.S. population in 1960 was 180 million and is projected to be 257 million in 1995.)

2.7. a. Using Table 2-3, determine the percentage of the total energy used for electrical power production which was used for hydropower in 1960.
 b. Same as part a but for the year 1995.

2.8. Same as Problem 2-7 except for nuclear power.

2.9. Using Table 2-2, determine the percentage of energy going to each of the four sections in 1960 and in 1995.

3

Electric Power Generation

3.1 IMPORTANCE OF ELECTRICAL ENERGY

Electric power is playing an ever increasing role in the use of energy in this country. It may be determined from Tables 2-1 and 2-3 that about 18 percent of the total energy used in 1980 was for electric power generation. It is projected that this figure will increase to 40 percent in 1995. On the absolute basis it is estimated that we will use about 5.6 times as much energy for electric power generation in 1995 as we used in 1960.

The rapid growth of electric generation in this country is shown in Tables 3-1 and 3-2, based on data in the *Annual Report to Congress, Volume 2,* by the Energy Information Administration, 1979. These tables show the growth in both the installed generating capacity and the electrical power produced by the various types of power plants. Among other things which these tables show is that although gas turbines constitute 8.5 percent of the total generating capacity, they produce only 1.2 percent of the electric energy which is generated, due to their low efficiency. Normally gas turbines are used for peak load conditions and for stand-by services. Several reasons account for the large growth in the demand for electrical energy:

1. It is convenient, readily available and is non-polluting for the user. (Pollution may be caused in its generation.)
2. The ever increasing use of machines, particularly in automation in our factories, requires much more electrical energy.
3. As the price of oil and gas continues to increase, it is anticipated that heat pumps will be economical to provide much of our building heating. (Heat pumps require electrical energy for their operation. See the Appendix for a discussion of heat pumps.)
4. If satisfactory electric storage batteries can be developed, it is predicted that electric cars will be used very extensively in the next decade or two.

Table 3-1. Installed Generating Capacity of Electric Utilities (Million Kilowatts)

	Conventional Steam	Hydropower	Internal Combustion	Gas Turbine	Nuclear	Geothermal	Total
1950	49.3	17.7	1.9	0	0	0	68.9
1955	87.1	25.0	2.4	0	0	0	114.5
1960	132.5	32.4	2.8	0	0.3	*	168.0
1965	186.6	43.8	3.4	1.4	0.9	*	236.1
1970	260.0	55.1	4.4	15.5	6.5	0.1	341.6
1975	352.9	65.9	5.1	44.1	39.8	0.6	503.8
1979	411.7	75.3	5.5	50.6	53.7	0.7	597.5

* Less than 0.05

Table 3-2. Production of Electrical Energy by Electric Utilities (Billion Kilowatt Hours)

	Conventional Steam	Hydropower	Internal Combustion	Gas Turbine	Nuclear	Geothermal	Total
1950	230	96	4	0	0	0	239
1955	430	113	4	0	0	0	547
1960	603	146	4	0	1	*	753
1965	851	194	5	1	4	*	1,055
1970	1,240	248	6	16	22	1	1,532
1975	1,414	300	6	22	173	3	1,918
1979	1,676	280	4	28	255	4	2,248

* Less than 0.5.

5. It is possible that large quantities of hydrogen will be produced by electrolysis of water at times when the loads on the power plants are low. This hydrogen could be used in place of natural gas, which is in short supply. Provided that satisfactory methods of handling it can be developed, hydrogen may be used extensively for automotive service.

Because of the growing importance of electrical energy, it is desirable to understand how it is produced and what forms of energy are required for its production. It is also desirable to understand the limitations and also the problems, particularly the environmental problems, associated with its production.

Before the start of the extensive hydroelectric power developments in the 1930's (T.V.A., Colorado River and Columbia River, for example), hydroelectric power plants supplied almost one-third of our electric power. Today, we obtain approximately 12 percent of our electric power from hydroelectric plants. Even though there has been a very large growth in hydroelectric power, the growth in our demand for electrical energy has been much greater. For example, in spite of the large amount of hydroelectric power produced by the Tennessee Valley Authority, the demand for electrical energy in that region is so great that over 75 percent of it is now produced in steam power plants.

Diesel engines are now used for power production in many small municipal power plants. The larger power companies use gas turbines to a limited extent, mainly for peak loads and for emergencies. However, approximately 80 percent of all electrical energy produced in this country is produced in steam power plants. Hence, the discussion in this chapter will be confined to the steam power plant.

Fossil fuels are the source of energy in most of our steam power plants but nuclear energy is becoming a more important source of energy. Geothermal energy is used in steam power plants for power production. Its use is expected to increase (see Chapter 14). The concept of the steam power plant is common to all types of plants, regardless of the source of energy.

3.2 THE STEAM POWER PLANT

Although some electric power is generated in industrial and municipal power plants, approximately 97 percent of the power generated in steam power plants is generated by the public utilities or by government power plants. Hence the discussion here will be centered on these types of power plants.

The simple steam power plant operates on what is known as the *Rankine cycle*. The elements of the Rankine cycle are shown in Figure 3-1. Steam is produced in the steam generator (boiler) at high pressure and temperature. In the turbine the steam expands to a very low pressure, producing work. The steam leaving the turbine enters the condenser where it is condensed, thus forming *condensate*. The pump draws the condensate from the condenser and builds up

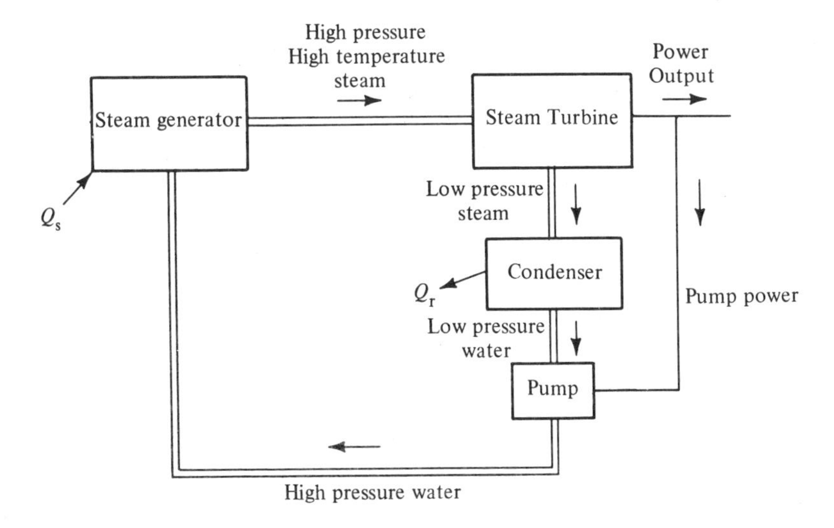

Figure 3-1. Simple steam power plant

sufficient pressure to force it into the steam generator. Because the pump feeds water into the steam generator, this water is known as *feedwater*. In the steam generator the water is turned into steam and then superheated (i.e., heated to a temperature in excess of the temperature at which it was formed).

The steam generator, as shown in Figure 3-1, is for a fossil fuel plant. For a nuclear power plant, a nuclear reactor furnishes heat to produce high pressure steam in a specially designed steam generator. The remainder of the system is the same as in Figure 3-1.

The steam turbine is the heart of the steam power plant. For fossil fuel power plants, the turbine receives steam at pressures normally ranging from 2,400 psia to 3,500 psia and at temperatures around 1,000°F. Upon entering the turbine the steam expands partially to a somewhat lower pressure in a row of nozzles, its enthalpy* being converted into kinetic energy. The steam, with velocities of perhaps 1,500 to 2,500 feet per second, is directed onto blades, thus imparting energy to them. The impact of the steam on the blades causes the wheel and the shaft to rotate. The blades are mounted on wheels which, in turn, are mounted on the shaft. After leaving the blades, the steam enters a second row of nozzles and undergoes further expansion. The kinetic energy developed by this expansion is transferred to blades mounted on a second wheel which is on the same shaft. This expansion process is continued, with 20 to 30 rows of nozzles, blades and wheels being required, depending on the steam conditions (see Figure 3-2). With the exception of the small amount of power required to overcome losses in the bearings, etc., the work delivered to the shaft by the wheels is delivered by the shaft of the turbine to the generator for production of electrical energy.

To obtain the maximum amount of work per pound of steam, it is expanded to the lowest feasible pressure. Frequently this pressure is as low as 0.5 psia. Such a low pressure is obtained by exhausting the steam into a condenser, where the steam is condensed by the circulation of water through the tubes of the condenser. The steam pressure in the condenser and hence at turbine exhaust is a function of the temprature of the circulating water. When the circulating water temperature is 40°F, the steam pressure at turbine exit may be as low as 0.25 psia. If, on the other hand, the temperature of the circulating water is 100°F, then the steam pressure at turbine exit may be about 1.7 psia. Although it may not appear to be the case, a small change in the turbine exhaust pressure will have a significant effect on the amount of work obtainable. This is illustrated in Table 3-3. In this table it is assumed that steam enters the turbine at 2,800 psi and 1,000°F. The theoretical work per pound of steam is shown in Table 3-3 for various exhaust pressures.

Table 3-3 shows why it is so desirable to use a condenser to produce as low a turbine exhaust pressure as feasible. By use of the condenser, with cold circulating water available, the theoretical work per pound of steam is about 40 percent greater than would have been obtained had the steam been exhausted

* See the Appendix for a discussion of enthalpy.

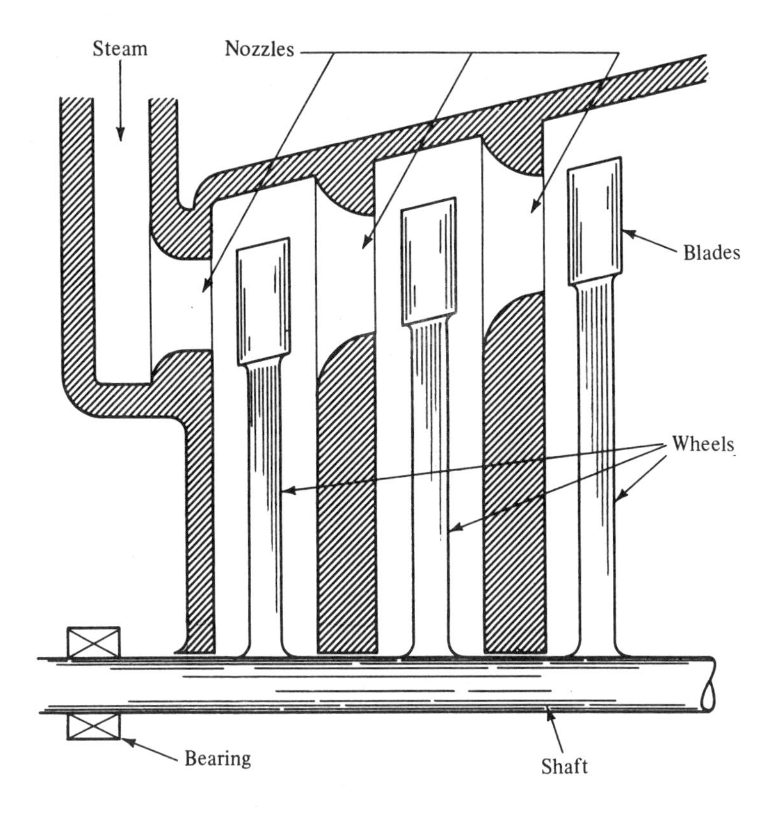

Figure 3-2. Elements of a steam turbine

Table 3-3. Theoretical Turbine Work

Exhaust Pressure, psia	Steam Temperature °F	Work, Btu per Pound of Steam
0.25	59.3	664.7
0.50	79.6	635.6
1.00	101.7	604.7
2.00	126.0	571.8
14.696	212.0	463.2

to the atmosphere. For a unit requiring 250 tons of coal to be burned per hour for an atmospheric exhaust turbine, there could be a saving of approximately 100 tons of coal per hour when a condenser is used. Assuming that coal cost 35 dollars per ton, this saving would be 3,500 dollars per hour or over 30 million dollars per year.

As the steam expands in the turbine, it has a tendency to partially condense. This condensation produces much moisture which can cause operating difficulties and will lower the efficiency of the turbines. To avoid excessive condensation and to improve the efficiency, after partial expansion, the steam is reheated to perhaps 1,000°F and then expanded to the condenser pressure.

The efficiency of a steam power plant can be improved by the use of feedwater heaters. Steam is drawn from that part of the turbine where the pressure is at the desired level and, as it condenses in the feedwater heater, heats the feedwater up to a temperature close to that of the condensing steam. Since the steam has already produced some work and since the latent heat* of the steam is fully utilized rather than wasted in the condenser, there is a gain in efficiency. In a large power plant the feedwater may pass through five or six feedwater heaters arranged in series, thus heating it up to within perhaps 100°F of the temperature at which it will boil in the steam generator. The details associated with all these heaters are too involved to be considered further in this text. It should be emphasized that the modern steam power plant contains not only these devices but many others to make certain that the maximum amount of work which is economically feasible is obtained from a given fuel input.

Figure 3-3 shows the elements of a simple condenser. A large condenser may contain as many as 80,000 to 100,000 tubes, one inch in diameter and perhaps 40 feet long. The circulating water passes through the tubes. The steam coming from the turbine surrounds the tubes. When it comes in contact with the cold surface of the tubes it condenses and drains to the bottom of the condenser. The condensate should be relatively free of minerals and dissolved gases and hence is pumped back into the steam generator where it is turned into steam.

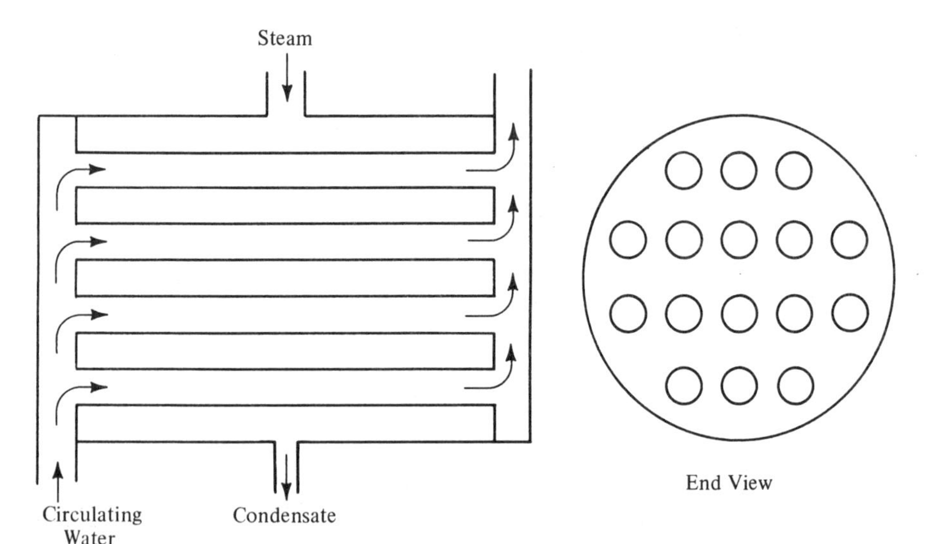

Figure 3-3. Elements of a simple condenser

* Latent heat is the heat given up by the steam as it condenses.

As discussed earlier, coal, oil, and gas are all used as energy sources for steam production. Of course the gaseous fuel is burned in the gaseous state. When oil is used it is atomized as it is forced into the furnace. Most of the oil is vaporized within the furnace. Formerly coal was burned largely on grates. Today it is common practice in large power plants to pulverize the coal before blowing it into the furnace. Since the fixed carbon* will not vaporize, combustion of it takes place at its surface. To prevent the coal particles from being swept out of the furnace by the hot gases resulting from combustion before it can be burned, the coal must be highly pulverized. Normally about two-thirds of the coal is reduced in size to approximately two-thirds of the diameter of a human hair.

The furnace of a steam generator of a modern power plant may have inside dimensions of 50 feet by 50 feet and 125 feet in height. Into such a furnace 250 tons of coal and 3,000 tons of air may be injected per hour. The walls of the furnace may be made up of tubes through which water is circulated. As the water receives heat from the burning fuel, it is vaporized, thus producing steam. The hot gases leaving the furnace are used both to superheat (i.e., heat it to above its boiling temperature) the steam and later to reheat it. As noted above, even though feedwater heaters are used, the feedwater enters the steam generator with a temperature below its boiling temperature. In a unit of the steam generator known as an economizer, the hot gases heat the feedwater up to its boiling temperature. Finally, the hot gases pass into an air preheater where they heat up the air to be used for combustion. By using these devices to abstract energy from the combustion gases, their temperature is reduced to about 275° to 325°F before they enter the chimney, thus using as much of their energy as is practical.

SUPPLEMENT A

ILLUSTRATIVE EXAMPLES

Example 3-1. Steam leaves a row of nozzles with a velocity of 2,000 feet per second at the rate of 4,000,000 pounds per hour. If 85 percent of the kinetic energy of the steam is delivered to the blade as work, determine the horsepower.

Solution. From the Appendix, the kinetic energy equals

$$\frac{mV^2}{2g_c}$$

where m is the mass, lb_m

V is the velocity, ft. per sec.

g_c is a conversion factor $= 32.174 \frac{lb_m ft}{lb_f sec^2}$

* See Chapter 6 for a discussion of the constituents of coal.

Then the kinetic energy per pound of steam is

$$\frac{1\times 2{,}000^2}{2\times 32.174}=62{,}160 \text{ ft. lb.}$$

The work given to the blades equals

$$62{,}160\times 0.85=52{,}840 \text{ ft. lb. per lb. steam}$$

The steam flow per minute is

$$\frac{4{,}000{,}000}{60}=66{,}670 \text{ lb. per min.}$$

The horsepower equals

$$\frac{66{,}670\times 52{,}840}{33{,}000}=106{,}700 \text{ hp} \qquad \text{answer}$$

Example 3-2. The density of air entering the furnace of a steam power plant is 0.043 lb. per cu. ft. Air is supplied at the rate of 3,000 tons per hour. Determine the volume of air supplied per second.

Solution: The airflow per second is

$$\frac{3{,}000\times 2{,}000}{3{,}600}=1{,}667 \text{ lb. per sec.}$$

$$\text{The volume rate of air flow}=\frac{\text{mass flow, lb. per sec.}}{\text{density, lb. per cu. ft.}}$$

$$=\frac{1{,}667}{0.043}=38{,}760 \text{ cu. ft./sec.} \qquad \text{answer}$$

Note: This air will occupy a cube 33.8 feet on a side.

Example 3-3. A steam turbine plant burns 320 tons of coal per hour. The heating value of the coal is 13,200 Btu per pound. The thermal efficiency of the plant is 39.8 percent. Determine the output of the plant in kilowatts.

Solution. Since

$$\eta_t=\frac{W}{Q_s},\ W=Q_s\times\eta_t.$$

$$\text{Then output}=\frac{320\times 2{,}000\times 13{,}200\times 0.398}{3413}=985{,}000 \text{ kW} \qquad \text{answer}$$

Example 3-4. Assume that 12 percent of the heat supplied by the coal of Example 3-3 is lost up the chimney and from the turbine and the generator.

Determine the amount of heat which must be given up in the condenser per second.

Solution. Of the heat supplied, 39.8 percent is delivered as work and 12 percent is lost. This totals 51.8 percent. The remainder, 48.2 percent, must be removed in the condenser. The heat supplied per second equals

$$\frac{320\times 2{,}000\times 13{,}200}{3{,}600}=23{,}470{,}000 \text{ Btu}$$

Using 2,000 lb. per ton and 3,600 sec. per hr.

The heat removed equals

$$23{,}470{,}000\times 0.482=1{,}131{,}000 \text{ Btu per sec.} \qquad \text{answer}$$

Example 3-5. The temperature rise of the circulating water which picks up the heat in the condenser of Example 3-4 is 15°F. Determine the tons of water required per second.

Solution: $Q=mc(t_2-t_1)$. Then

$$m=\frac{1{,}131{,}000}{1\times 15\times 2{,}000}=37.7 \text{ tons per second}$$

Note: at these low temperatures c for water is 1.0 Btu/lb.-°F.

SUPPLEMENT B

PROBLEMS

3-1. Steam leaves a row of nozzles with a velocity of 1,800 feet per second. 84 percent of the kinetic energy of the steam is delivered to the blade as work. The blade horsepower is 92,000. Determine the steam flow in pounds per hour.

3-2. The air in Example 3-2 flows through a square duct with a velocity of 110 ft. per sec. Determine the size of the duct, in feet. Note that the product of the cross section area and the velocity equals the volume rate of flow.

3-3. A steam power plant burns 300 tons of coal per hour when the turbine exhaust pressure is 0.50 psia. Estimate the amount of coal required when the exhaust pressure is 2.00 psia. Assume that the amount of coal required is inversely proportional to the theoretical work. See Table 3-3.

3-4. Assume that the temperature of the circulating water leaving the condenser is 5 degrees lower than the steam temperature, comment on the feasibility and the desirability of using this water for heating purposes when the turbine exhaust pressure is 0.5 psia and also when it is 2.0 psia.

3-5. There are 80,000 tubes in a condenser. The tube length is 40 feet. Determine the total length of the tubes in miles.

3-6. Determine the cubic feet of water required in Example 3-5. Assume that the density of water is 62.4 lb. per cu. ft.

3-7. Assume that the work delivered per pound of steam in Example 3-3 is 575 Btu. Determine the tons of steam which must be produced per hour.

3-8. Refer to Problem 3-6. Assume that a cooling tower is to be used to cool the circulating water. Assume that 1½ percent of the water entering the tower is lost by evaporation and drift (water droplets carried away by the air). Determine the gallons of make-up water required per day (24 hours). (There are 8.34 lb. per gal.)

3-9. List major factors to be considered in locating a coal burning steam power plant.

3-10. The power plant in Example 3-3 is considering burning oil, having a heating value of 142,000 Btu per gallon. Assuming the same efficiency, determine the number of barrels of oil required per hour. (42 gallons per barrel.)

4

Economics of Energy Production

Technically it is feasible to obtain energy from many sources. At present, however, most of these sources are not utilized because it is not economically sound to do so. It may be argued that in this time of ever growing energy shortages we should disregard the cost of procuring energy, the cost consideration being a short-coming in our capitalistic society. However, it is essential to ask why it is too expensive to use certain methods of using particular energy sources. The reasons that the more efficient systems may be more expensive initially than the present systems are

1. The proposed method requires the use of scarce, high price materials.
2. Execessive amounts of energy are required to produce the equipment for the proposed method.
3. Much manpower is required.

As one example, consider the early use of solar cells for electric power production. They were extremely expensive since the silicon had to be highly refined and much direct labor was required in the production. Furthermore, so much energy was required in the manufacture of the solar cell that it would have to be operated for 50 years to get back the energy expended in its manufacture. Obviously no society, communist or capitalist, could be justified in using these cells for power generation except in special cases, such as spacecraft use. Improvements in the methods of manufacture of solar cells have materially reduced the cost of manufacture but the present costs are still much too high to justify their use for bulk power generation.

As a second example, consider electric power plants. It was pointed out in Chapter 3 that the steam temperature entering the steam turbine is approximately 1,000°F. We do have materials that will permit operation at least as high as 1,200°F. Operating at 1,200°F would be materially more efficient and hence would give a significant reduction in the amount of fuel required. We do not use steam temperatures of 1,200°F since very expensive alloy steels must be used to stand this temperature. We must import many of the materials to produce these steels, such as chromium, manganese, and nickel. Worldwide, they are in short supply and therefore are expensive.

4.1 FIXED COSTS

Two types of costs are involved in the production of any product: the fixed costs and the operating costs. As the name implies, *fixed cost* is the cost which must be paid for simply owning a piece of equipment. It is independent of whether the equipment may produce at its maximum capacity or it may not be used at all.

There are four factors involved in fixed costs:

1. The initial or capital cost of the equipment
2. Interest on the money required for the initial cost
3. Taxes*
4. Insurance

Consider, for example, the fixed costs of solar heating a house. Unless a very large and expensive solar heating system is installed, a full size conventional heating system will be required to supply heat when several cold dark days occur in a row. Thus the solar heating system is in addition to the conventional system and does not supplant it. The owner may feel that the installed cost of a solar heating system for his house will be $8,000. (It may materially exceed this figure before it is completed.) The owner feels that his system will last him for at least 20 years. Accordingly, he divides the $8,000 by 20 and concludes that it will cost him $400 a year to heat his house. Since it is now costing him $700 a year for heat, he concludes that he will save at least $300 a year (or more as the price of fuel increases). Careful calculations, however, will show that he will lose money. Probably his conventional system will use $200 worth of fuel during the periods of cold dark days. He has neglected the cost of maintenance on his solar heating system, which probably will not be completely offset by the reduction in maintenance cost of his conventional system. He has neglected interest costs, taxes, and insurance. Although it is possible that the federal government will make low interest loans available for this purpose and also may minimize taxes on solar heating systems, it should be evident that solar heating systems cannot be economical unless the original cost can be reduced significantly.

The effect of fixed cost factors can be illustrated by considering the desirability of installing a solar heating system in a large building. Assume that the present cost of heating the building is $22,000 per year. This cost may be reduced to $3,000 per year if a solar heating system is installed. Let us make the following assumptions:

Initial cost of the system	$210,000
Useful life of the system	20 years

* As used here, we are concerned here with the net taxes after depreciation allowances and other tax write-offs have been made.

Interest	12 percent
Taxes	1 percent
Insurance	½ percent
Salvage value*	15 percent

There are various methods of obtaining the money required for the initial purchase. For simplicity, assume that bonds will be issued which will be paid off at the end of 20 years. Each year money will be placed in a sinking fund where it will draw interest. There must be sufficient money in the sinking fund so that when added to the salvage value, there will be the $210,000 available to pay off the bonds. The salvage value of the plant equals 210,000×0.15=$31,500. Thus the amount to be placed in the sinking fund is $210,000−$31,500=$178,500. The amount of money to be placed in the sinking fund each year can be determined by use of the sinking fund factor.

$$\text{Sinking fund factor} = \frac{r}{(1+r)^n - 1} \tag{4-1}$$

Where r is the sinking fund interest rate
n is the number of years

Assuming that the money placed in the sinking fund can draw 8 percent interest, our sinking fund factor equals 0.0219. Then the yearly payments into the sinking fund equals 178,500×0.0219=$3,909.

The total fixed charges per year are:

Sinking fund payments	$ 3,909
Interest: 0.12×$210,000	25,200
Taxes: 0.01×$210,000	2,100
Insurance: 0.005×$210,000	1,050
Total	$32,259

The reduction in the operating costs† for heating the building is $22,000−$3,000=$19,000. By subtracting the $19,000 from the fixed charges of $32,259, we find that it costs $13,259 a year to use the solar system. If we had disregarded other factors and simply divided the original cost minus the salvage value ($178,500) by the 20 years, we would have a yearly cost of $8,925. Subtracting this figure from the savings in operating costs of $19,000 we would conclude, erroneously, that we would save $10,075 per year.

Assume that the government were to provide loans at very low interest rates (5 percent) and forgive taxes, the fixed charges per year in this case would be $14,549. Subtracting this figure from the savings in the operating costs of $19,000 we find that we will save $3,541 per year by installing the solar system. However, recognize that this savings is possible only because the taxpayers pay for the difference between the actual loss of $13,259 per year and the revised savings of $3,541.

* Salvage value is the percentage of the original cost which can be obtained by selling the plant as scrap at the end of its useful life.

† Operating costs are the total of all costs for operating a system.

It is not possible to predict accurately the total costs of using or not using a solar heating system over a 20 year period. Fuel costs certainly will increase, as will the initial cost of the solar system. This example does show, however, that there are many important factors which affect the desirability of installing a solar heating system.

It cannot be emphasized too strongly the need for predicting the initial cost of any system as accurately as possible, particularly if the system is an expensive one. When the system under consideration is similar in most respects to existing systems and when the installation is to be completed within a year or two, it should be possible to make a close estimation of the initial cost, making due allowance for inflation. If, however, the system is not to be installed for five or ten years, it is difficult to predict accurately what the final cost of the system will be. This is particularly true for power plants. There are several reasons for the uncertainty in predicting the cost of a power plant scheduled to go into operation in eight to ten years after a decision is made to install it. Some of these reasons are

1. Inflation.
2. Probable addition of equipment required to meet new environmental standards.
3. Uncertainty as to how long it will take to build the plant and hence uncertainty on the interest payments during plant construction.

The initial cost of a nuclear power plant is somewhat larger than a coal fired plant. But the extra fixed charges for a nuclear plant are more than offset by its much lower fuel costs. As a result, at present power can be produced for a significantly lower cost in a nuclear plant. How about future costs in these plants?

The initial costs of both fossil fuel and nuclear power plants are rising rapidly. In the 1960's, the initial cost of a large fossil fuel power plant was about $100 per installed kilowatt capacity and one nuclear plant had an initial cost of $135 per kilowatt. Present costs are at least nine times these figures and are rising.

There are several reasons for the rapid increase in the costs of power plants over a period of years. Inflation is only one of these reasons. Steam pressures and temperatures have increased. Equipment has been incorporated in the plant to improve the efficiency, thus adding significantly to the cost of the plant. New environmental controls have added greatly to the cost and will probably continue to increase. Because it takes so much longer to construct a plant now, there is much more interest to be paid on the plant as it is being constructed. One estimate of the initial cost of a nuclear plant in 1985 is $1,300 per kW and $900 per kW for a coal fired plant. It should be noted that five to six years are required to bring a coal fired plant on line after a decision is made to install the plant. Approximately eleven years are required in the case of a nuclear power plant. Much of this time is required because of the many hearings which must be held and because approval must be obtained from at least

50 bureaus and agencies. In spite of so many uncertainties as to the magnitude of fixed charges, power companies' officials must make decisions now as to whether coal fired or nuclear plants will produce power more cheaply ten years from now.

When new systems are to be installed, estimates of their initial costs are extremely difficult to make. This is particularly true when an energy producing system has been scaled up from a laboratory model to a full size plant. In many cases, there have been cost overruns of several hundred percent because many factors are encountered in the full scale plant that were not present in the laboratory model. It is to be expected that the initial costs will be reduced as more full scale plants are built but it is very difficult to predict how much the reduction will be. And yet, with all these uncertainties, our business and government leaders are called upon to make decisions about investing in alternative energy systems.

4.2 OPERATING COSTS

Operating costs are closely related to the extent of operation of a given system. If the system is not operated, normally the operating costs are very low with only preventive maintenance required. However, in some cases, skeleton crews may be kept on the job when shut-downs are for short periods.

In general, operating costs can be broken down into the following categories:

1. Energy costs (fuel and electrical energy).
2. Supplies.
3. Maintenance and repairs.
4. Labor.

The first three items vary almost directly with the output of the plant as will labor costs for piece-workers. This is only partially true for office workers and for the supervisory personnel.

As with fixed charges, it is relatively easy to estimate the operating costs with a fair degree of accuracy for the near future for plants similar to those now in existence, provided that it is possible to estimate rather accurately the percent capacity at which the plant will be operated. It is much more difficult to predict the operating costs for a new type of plant.

4.3 SUMMARY ON ECONOMICS OF ENERGY PRODUCTION

If we try to use alternative energy producing systems without considering the economics of the systems, we may use more energy in making them than we can gain from the systems. We may also waste both scarce materials and labor.

In considering the economics of the systems, we must consider all phases of fixed charges, including interest, return of investment, taxes and insurance. In predicting the operating costs we should try as carefully as possible to specify the percentage of the maximum output at which the systems will actually operate. Arriving at this percentage, we should then estimate energy costs for operation, supplies, maintenance and repairs, and labor costs. To obtain the total cost of our product we should add the fixed charges to the operating costs. These total costs should be compared with the present energy production costs of existing systems as well as with predicted future costs.

There is a great uncertainty of predicting both the fixed and operating costs for new types of systems, particularly large ones. Hence, even though it may be very time consuming, it is desirable to build pilot plants at first, not only to demonstrate the technological feasibility of various energy producing methods but also to gain sufficient data to predict with some assurance the economic feasibility of larger plants.

SUPPLEMENT A

ILLUSTRATIVE EXAMPLES

Example 4-1. A power plant rated at 50,000 kW has an initial cost of $800 per kW. Assume the plant capacity factor (i.e., the ratio of the average load to the rated load) is 45 percent. Assume a useful life of 20 years and a salvage value of 14 percent. Interest on the bonds will be 12 percent and for the sinking fund 8 percent. Taxes and insurance total 2¼ percent.

Determine:

a. Fixed charges per year.
b. Fixed charges per kWh produced.

Solution. Initial cost of the plant = 50,000 × 800 = $40,000,000.

$$\text{Sinking fund factor} = \frac{0.08}{(1.08)^{20} - 1} = 0.0219$$

Sinking fund accumultation = 40,000,000 − 0.14(40,000,000) = $34,400,000

(a)

Interest = $40,000,000 × 12 =	$4,800,000
Sinking fund payment = $34,400,000 × 0.0219 =	753,400
Taxes and insurance = $40,000,000 × 0.0225 =	900,000
	$6,453,000 answer

(b) Energy produced per year $= 50{,}000 \times 0.45 \times 8{,}760$
$= 1.971 \times 10^8$ kWh per yr.

where 8,760 equals the hours per year. The fixed cost per kWh =

$$\frac{6{,}453{,}000 \times 100}{1.971 \times 10^8} = 3.274¢/\text{kWh} \qquad \text{answer}$$

Note that this is the fixed cost only and does not include the operating costs nor the cost of delivering the electricity to the customer.

Example 4-2. The power plant in Example 4-1 uses oil which costs 75¢ per gallon. It produces 13.2 kWh per gallon of fuel. Labor costs per year total $420,000. Maintenance, repairs and supplies cost 1.6 mills per kWh.

Determine the operating cost and the total cost of producing power per kWh. Note: a mill is a tenth of a cent.

Solution. The amount of fuel required per year is

$$\frac{1.971\times 10^{8}}{13.2}=14{,}932{,}000 \text{ gal. per year}$$

$$\begin{aligned} \text{Fuel costs} &= 14{,}932{,}000\times 0.75 = \$11{,}200{,}000 \\ \text{Maintenance, etc.} &= 1.971\times 10^{8}\times 0.0016 = 315{,}000 \\ \text{Labor} &= 420{,}000 \\ \text{Total} &= \$11{,}935{,}000 \end{aligned}$$

$$\text{Operating cost per kWh}=\frac{11{,}935{,}000\times 100}{1.971\times 10^{8}}=6.055¢/\text{kWh} \qquad \text{answer}$$

$$\text{Total cost}=6.055+3.274=9.329¢/\text{kWh} \qquad \text{answer}$$

Example 4-3. Two customers have a maximum demand of 10,000 kW. Customer A uses 20,000,000 kWh per year and customer B uses 50,000,000 kWh per year. The initial cost of the power plant and distributing system is $700 per kW. Total fixed charges are 16 percent. The operating cost is 2.5¢/kWh. Determine the cost of power for each customer, cents/kWh.

Solution. Customer A:

$$\begin{aligned} \text{Fixed costs} &= 10{,}000\times 700\times 0.16 = \$1{,}120{,}000 \\ \text{Operating costs} &= 20{,}000{,}000\times 0.025 = 500{,}000 \\ \text{Total} &= \$1{,}620{,}000 \text{ per year} \end{aligned}$$

$$\frac{1{,}620{,}000\times 100}{20{,}000{,}000}=8.10¢/\text{kWh} \qquad \text{answer}$$

Customer B:

$$\begin{aligned} \text{Fixed costs} &= 10{,}000\times 700\times 0.16 = \$1{,}120{,}000 \\ \text{Operating costs} &= 50{,}000{,}000\times 0.025 = 1{,}250{,}000 \\ \text{Total} &= \$2{,}370{,}000 \text{ per year} \end{aligned}$$

$$\frac{2{,}370{,}000\times 100}{50{,}000{,}000}=4.74¢/\text{kWh} \qquad \text{answer}$$

SUPPLEMENT B

PROBLEMS

4-1. Consider using a much cheaper solar heating system for the very large building discussed in the text. This system will have a useful life of 10 years but will cost only $160,000. Assume that the other factors are the same as those used in the text. Determine the annual fixed charges.

4-2. Same as Problem 4-1 except that the government will provide a loan at 3½ percent interest and the system will be tax exempt.

4-3. Same as Example 4-1 except the plant capacity factor is 70 percent. Compare the results and comment on them.

4-4. Same as Example 4-2 except the plant capacity factor is 70 percent. Compare the results and comment on them.

4-5. Refer to Example 4-3. Since customer B demands much more power, which requires much more fuel for its generation, should customer B be charged more per kWh to discourage him from using as much power?

4-6. A family takes a vacation for a month. They feel that they should not receive an electricity bill for that month since they used no electrical energy. Make a comment.

4-7. Some people feel that industrial customers should be charged the same cost per kilowatt hour as residential customers. Make a comment.

4-8. With rising costs of electrical energy, very poor people are finding it difficult to pay their electric bills. Should very poor people be given special low rates?

PART II

OUR PRESENT ENERGY RESOURCES

5

Introduction

As was discussed in Chapter 2, our present demands for energy are met largely by the use of fossil fuels, primarily oil and natural gas. Hydropower does account for approximately 4 percent of our energy needs and nuclear power approximately 1 percent, with the percentage supplied by nuclear power increasing. The next five chapters of Part II will examine the extent of these present resources and discuss the probability of their meeting our future demands.

The reported amounts of remaining resources vary widely. For example, some reports indicate that our domestic oil supply will be pretty well exhausted in twenty years. Other reports show that we have sufficient oil reserves in this country to last for at least fifty years. It should be recognized that it is almost impossible to ascertain with a high degree of certainty the total amount of any mineral, particularly at depths of several miles and off shore beneath the oceans. Some tabulations of the amounts of mineral resources do not specify whether or not the amounts given are based on known evidence or are purely speculative, whereas other tabulations break down the amount of resources into two or three classifications. One such method of classification lists (1) proved* reserves, (2) probable reserves, and (3) possible reserves.

The proved reserves are determined by means of test borings made in a systematic way throughout a known deposit. The proved reserves are thought to be accurate within plus or minus 25 percent. The probable reserves generally are located adjacent to the proved reserves in geological formations similar to that of the proved reserves. There is no real evidence that the possible reserves exist other than that the geological formation in the given area is such that it is possible for the given mineral to be present.

Some statements have been made that the large oil companies do know how much oil there is in their own fields but are concealing it from the government in order to profit from an apparent shortage. If the oil companies do know

* The term *proven* is also used.

where oil exists, we must ask why so many oil and gas wells which are drilled turn out to be dry. In 1979, a total of 49,740 oil and gas wells were drilled. Of this number, 15,740 or 31.6 percent were dry. The average cost of drilling each dry well was approximately $300,000. (The average depth of the dry wells was about one mile.) Of all the wells drilled, 10,510 were exploratory wells. Of the exploratory wells drilled, 7,490 were dry (i.e., 71.2 percent).

There are two factors which decrease the amount of a fossil fuel that is presently recoverable. In many cases, it is not economically feasible to mine a given mineral, using existing technology. For example, test borings may show the existence of a coal seam a few feet thick at a depth of several thousand feet. Obviously, the present price of coal is too low to make it desirable to mine this coal. It is not possible to predict whether or not new mining techniques will be developed and/or coal prices will so increase that it will be desirable to mine this coal some time in the future.

A second factor affecting the amount of a given mineral which can be delivered for use is the inability to recover all of the mineral existing in the earth. For example, present practice is to leave pillars of coal behind for mine roof support as the coal is mined. Oil frequently is locked up in pockets which may or may not be readily broken to obtain the oil. Some oil may be so viscous that it will not flow into the oil well unless heated.

It should be evident from the above discussion that it is very difficult to state with certainty the amount of fossil and nuclear fuels still remaining in the earth, particularly the amount available for our use. It is possible to calculate fairly accurately the amount of hydropower which can possibly be developed in the future. It is not possible, however, to evaluate the extent of environmental concerns which will prevent the full development of the hydropower potential.

Because it is so necessary to have some concept of the total amount of our present energy resources, the next five chapters will present what are believed to be the most reliable estimates of the useable amounts of coal, oil, natural gas, uranium and hydropower which are available for our use.

6

Coal

6.1 THE NATURE OF COAL

Coal may be described as a combustible solid formed from fossilized plants. Eons ago, apparently, large accumulations occurred of trees, grass, shrubs, and swamp vegetation. At first, there may have been biochemical decay brought about by action of bacteria. Later, the vegetation was buried by sediment, preventing any further decay. These deposits were acted upon by the pressure of the covering layer and by heat, which changed the structure and composition of the vegetation and produced what we know as coal.

In the transformation of the vegetation, peat was formed first as some of the readily volatized portion of the vegetation was driven off. In general, peat has a woody structure and, particularly when found in bogs, its moisture content is very high. Although it is not a coal, it can be used as a fuel after it has been dried. Assuming that the peat is acted upon continuously by pressure and heat, its structure is changed and it is transformed into first *lignite coal,* then *subbituminous coal,* next *bituminous coal,* and finally into *anthracite coal.* During the transformation process, more and more of the volatile portions of the coal, including moisture, are driven off. The total transformation process is a very lengthy one and much of our coal has not been subjected to the combination of heat and pressure sufficiently long to produce anthracite coal.

One method of specifying the constituents of the coal is the *proximate analysis*. This analysis reports four constituents: 1) moisture, 2) volatile matter, 3) fixed carbon, and 4) ash. The general specifications for obtaining these constituents are as follows. A representative sample of the coal is crushed and heated to approximately 220°F for an hour. The decrease in weight due to this heating is said to be the *moisture* content of the coal. If the coal is further heated to approximately 1,740°F for seven minutes, in the absence of air, the *volatile* constituents will be driven off. These are largely hydrocarbons, ranging from methane to tars. The *fixed carbon* is not driven off by the application of heat but will burn when heated to a high temperature in the presence of oxygen. The *ash* is the inert matter, which will not vaporize under normal heating conditions and which will not burn.

A very important property of coal is the heating value. Heating values of fossil fuels were discussed in Chapter 1.

Most lignite coals are high in both ash and moisture, and anthracite coals generally are low in both. There are, however, considerable variations in both ash and moisture and a given type of coal. Hence, the amount of volatile matter and fixed carbon in the *combustible* part of the coal is used to classify the various types. (The combustible is composed of the volatile matter and the fixed carbon.)

A second method of specifying the constituents of a coal is known as the *ultimate analysis*. This analysis reports the amounts of carbon, oxygen, hydrogen, nitrogen, and sulfur in the coal. The amount of ash in the coal is also reported. The ultimate analysis is a chemical analysis. As such, it requires much equipment and the services of a trained chemist. On the other hand, the proximate analysis of a coal is much easier to determine.

There is no full agreement as to the method of classification of coals. In general, however, the combustible part of lignite is at least 45 percent volatile matter, subbituminous coal combustible is at least 40 percent volatile, bituminous coal combustible is at least 15 percent. The volatile matter in the combustible of anthracite coal ranges from 1 to 14 percent.

Heating values of coal may be reported for the coal as it exists (as received), on the moisture free basis, or on the moisture and/or ash free basis. The heating value of the coal, as received, is of prime interest to the consumer since the coal which he buys and for which he pays transportation costs includes the moisture and the ash.

Because of the wide range in the amount of moisture and ash in coals, particularly in the lignite coals, there is a wide range in the heating values, even for one type of coal. A range of heating values for various types of coal is given in Table 6-1. It is possible that a few coals may be outside the range given in Table 6-1.

Table 6-1. Heating Values

Type of Coal	Heating Value, as Received Basis, Btu per lb.
Lignite	5,700 to 7,500
Subbituminous	8,300 to 10,000
Bituminous	12,000 to 14,800
Anthracite	13,000 to 14,500

Before examining the extent of our coal reserves, one more factor should be noted. That is the sulfur content of the coal. When the coal is burned, substantially all the sulfur forms sulfur dioxide which is environmentally objectionable. Environmental Protection Agency (EPA) regulations prohibit the burning of coal containing more than 1 percent (in some cases, particularly for the lignite and subbituminous coals, less than 0.5 to 0.8 percent) sulfur unless precautions are taken to reduce the sulfur dioxide effluents to acceptable levels. Almost all of the anthracite coals, and most of the subbituminous coals and the lignite coals contain less sulfur than the maximum allowable amount. On the other hand, between 80 and 85 percent of the bituminous coals lying east of the Mississippi River contain excessive amounts of sulfur.

6.2 OUR COAL RESOURCES

Coal is by far the most abundant of our fossil fuels. Coal is known to exist in 32 states. It has been estimated that some 350,000 square miles or about 11 percent of our total continental U.S. land area is underlaid with coal. In many cases, however, coal exists in thin veins and frequently at such depths that it is not presently economical to mine it.

As discussed earlier, it is very difficult to estimate closely the amount of any mineral existing below the surface of the earth. Thus there is considerable disagreement in the reported reserves of coal. The U.S. Bureau of Mines periodically makes a very thorough study of the coal reserves in this country. Based on such a study, Table 6-2 gives the amount of reserves by type of coal, its location (east or west) and whether it can be mined by surface operation or underground. Table 6-3 is similar but it lists the sulfur content of these coals.

Table 6-2. U.S. Coal Reserves (Billions of Tons)

	Anthracite	Bituminous	Subbituminous	Lignite	Total
Underground					
East of Mississippi River	7	155	0	0	162
West of Mississippi River	*	27	108	0	135
Total	7	182	108	0	297
Surface					
East of Mississippi River	*	39	0	1	41
West of Mississippi River	*	8	61	33	101
Total	*	47	61	34	141
Grand Total	7	229	168	34	438

* Less than 0.5 billion tons.

"Demonstrated Coal Reserve Base by Rank, Potential Method of Mining and Region, January 1, 1976." *Annual Report to Congress*, Volume II, 1979, p. 123. Energy Information Administration, U.S. Department of Energy, Washington, D.C.

It is predicted that at present, it is economically feasible to mine about 50 percent of these coal reserves or somewhat over 200 billion tons. The amount of coal mined in this country in 1979 was 747 million tons, of which 66 million tons were exported. Thus, if we continue to mine coal at the present rate, we

Table 6-3. Sulfur Content of U.S. Coal Reserves (Billions of Tons)

	Sulfur Range				
	1% or Less	1.1 to 3%	Greater Than 3%	Unknown	Total
Underground					
East of Mississippi River	26	47	64	26	162
West of Mississippi River	102	11	8	14	135
Total	129	58	71	39	297
Surface					
East of Mississippi River	7	8	19	7	41
West of Mississippi River	66	26	3	5	101
Total	73	34	22	12	141
Grand Total	201	92	94	51	438

"Demonstrated Coal Reserve Base by Sulfur Content, Potential Method of Mining, and Region, January 1, 1976." *Annual Report to Congress,* Volume II, 1979, p. 125. Energy Information Administration, U.S. Department of Energy, Washington, D.C.

can do so for 275 years. However, it may become economically feasible to mine more coal. It has been estimated by the U.S. Geological Survey that the total identified coal resources in this country total 1,600 billion tons and that there are possibly another 1,600 billion tons as yet unidentified. As energy prices rise and as mining methods are improved, it is expected that a considerable amount of this coal will be mined. It is evident that we have vast coal resources, some 30 percent of the known, presently recoverable coal reserves in the world. Thus it may seem that coal is the solution of our energy problem.

However, the problem is far from being simple. At present, coal supplies less than one-fifth of our energy needs. As other energy sources become depleted, it is expected that coal will meet an ever increasing percentage of our energy needs. This fact, coupled with our increasing energy demands, has led to the prediction that the use of coal at the end of the next two decades will be over twice the present consumption. Of course, if our use of coal continues to increase at such a rate, coal will be available to us for a much shorter period of time than indicated above.

The second factor involved in the ability to solve our energy problem by using coal is embodied in the statement of S. David Freeman.* "There are two things wrong with coal today. We can't mine it and we can't burn it."

6.3 MINING COAL

Let us examine Freeman's statement. At present, approximately one-half of our coal is obtained by strip mining. In the strip mining of coal, the over-

* S. David Freeman, as quoted in Hammond, Metz and Maugh, *Energy and the Future,* American Association for the Advancement of Science, 1973.

burden is stripped off the coal seam, leaving the coal ready to be scooped up by giant power shovels. In the years past, many strip mine operators have ravaged the landscape in their efforts to get at the coal. The overburden was piled more or less indiscriminately. Particularly in hilly country, rains washed the overburden down across farms and other private property into streams, thus wreaking havoc in the region. More recently, however, some strip miners have been very careful after the coal has been removed to restore the land to at least as good a condition as before mining.

It is expensive in hilly regions to retain the overburden for refill. Under some conditions, it has been stated that the cost of restoring land will be at least $6,000 an acre. An increasing amount of strip mining is of lignite and subbituminous coals. Most of these coals are located in Montana, Wyoming and the western Dakotas. This is a dry region in which rainfall is insufficient during much of the year to initiate and maintain the desired vegetation on the restored land. Although some water may be withdrawn from rivers for this purpose, this water supply is limited. Because of the difficulty of restoring the land to an acceptable condition after the strip mining operations, bills to restrict strip mining operations have been introduced in some state legislatures and have been passed by Congress. The bills passed by Congress were vetoed by President Ford on the grounds that they were too restrictive. On the other hand, many feel that it is essential to produce this coal to help meet our energy needs, even if it may be necessary to sacrifice some of our environmental goals.

Underground coal mining is a hazardous occupation. There is danger of mine fires and explosions. Methane gas normally is present in coal beds. Methane is highly explosive and hence extreme care must be used to prevent explosions. Some consideration is being given to pumping the methane from a coal bed before mining it. In this way, not only are the dangers of fires and explosions reduced, but in certain cases it is possible to collect the methane in sufficient amounts so that it is economical to sell it. Unless the mine roof is supported properly at all times, there is danger of cave-ins. In addition, because of the large amount of dust present, many miners contract black lung disease. To promote mine safety, federal regulations of coal mines are becoming much more stringent, although not as stringent as the miners desire. These safety requirements have greatly increased the cost of mining coal and have driven some of the smaller mine operators out of business.

Generally, the water pumped from mines is acid or otherwise contaminated. Formerly this water was allowed to flow into neighboring streams, occasionally with disastrous results. It is common practice to wash the coal after it is brought to the surface to remove some of the slate and other ash. In the past, this waste material was placed indiscriminately in enormous piles, creating a desolated looking region. Although costly, it is now recognized that it is necessary to control the adverse effects of mine water and the discarded ash.

6.4 BURNING COAL

Unless burned properly, coal will cause much pollution. In the smaller installations, coal is burned on grates. They must be designed properly for the particular type of coal being burned and unless operated properly, combustion will not be complete. Carbon monoxide and some hydrocarbons as well as smoke may be discharged, thus polluting the atmosphere. Most of the coal burned in the large public utility power plants is in pulverized form. The coal is so highly pulverized that about 65 percent of it has a diameter less than about two-thirds of that of a human hair. By careful control of the degree of pulverization and of the amount and the mixing of air with the coal, no significant amount of hydrocarbons are emitted from the furnace and the amount of carbon monoxide is very small. Because of the high temperatures involved, nitrogen oxides are formed. It is recognized that these oxides may be a health hazard, and hence it may be necessary to establish more stringent standards, necessitating a change in the method of combustion to meet these standards.

Generally the ash in the gases leaving a pulverized coal furnace has been solidified. Much of this ash will be carried up the chimney unless methods are taken to remove it from the gas stream. Tall chimneys help to disperse the ash (called particulate matter). However, since the large power plants burn several hundred tons of coal per hour, the amount of particulate matter falling on the area in the vicinity of the plant becomes excessive even when tall stacks are used. Various methods have been used to remove most of the particulate matter, but many of these methods cannot remove a sufficiently large percentage to meet present standards. The trend today is toward the use of electrostatic precipitators in which the particulate matter is charged electrically and is then attracted to charged plates. Under favorable conditions, electrostatic precipitators can have an efficiency of over 99 percent (i.e., they remove over 99 percent of the particulate matter).

Most of the coal burned in this country is burned east of the Mississippi River. Approximately one-sixth of all coal burned in this country is burned in three states: Indiana, Ohio, and Pennsylvania. With the exception of the anthracite coal in eastern Pennsylvania, most of the coal east of the Mississippi River is bituminous coal. In many respects it is an excellent coal and it is ample in supply. Generally, it is low in ash and moisture. Its heating value is high. Unfortunately, much of it contains excessive amounts of sulfur. Whether it is burned on grates or in the pulverized form, substantially all of the sulfur in the coal is discharged as sulfur dioxide. In the atmosphere, much of the sulfur dioxide turns into sulfur trioxide and then into sulfuric acid. Thus, the discharging of sulfur dioxide into the atmosphere has a detrimental effect on humans, on plant life, and on physical objects.

There are so many variables involved that it is very difficult to establish reasonable standards applicable to all conditions as to the allowable amount of sulfur dioxide which may be discharged into the atmosphere. However, the Environmental Protection Agency (EPA) has established such standards. It is not easy to justify these standards under some conditions. However, many environmentalists feel that the standards are too lenient. In fact, some localities have established standards more stringent than those of the EPA. It should be noted that some agronomists feel that there should be some sulfur compounds in the air in remote areas since it is desirable to supply the sulfur required for healthy plant growth. It must be recognized, however, that it is very difficult to establish a standard which properly protects the environment without unduly preventing the burning of much of our coal.

Unfortunately, over 80 percent of the bituminous coal contains too much sulfur to meet the EPA standards unless control methods are used. Although, as stated earlier, most of our subbituminous and lignite coals contain a sufficiently small amount of sulfur to meet EPA standards, these coals are far removed from the eastern part of the country where most of the coal is needed. The high ash and moisture content of the subbituminous and lignite coals, together with their low heating values, means that larger quantities of these coals must be used, thereby increasing the cost of transportation over long distances. Thus, if coal is to be used in increased amounts in the eastern part of the country, methods must be developed to control the sulfur dioxide discharged when burning bituminous coals.

An ever increasing amount of effort is being devoted to controlling sulfur dioxide discharges. These efforts have been concerned with all three possible areas of control: (1) before combustion, (2) during combustion, and (3) after combustion. Hundreds of methods of controlling sulfur dioxide have been proposed. Many of these have been demonstrated successfully in laboratory tests. In general, however, no one method is completely satisfactory as yet for application to large power plants.

Before combustion, some sulfur may be removed by carefully controlled washing. In addition, the coal may be treated chemically to remove the sulfur. It may also be solvent extracted or gasified, with the sulfur being removed during these processes. See Chapter 20 for a discussion of these two processes.

For treatment during combustion, a substance such as limestone may be injected into the furnace. This is the most simple method to reduce sulfur dioxide formation. Lime or pulverized limestone is blown into the furnace to react with the sulfur in the coal. (It is assumed here that the coal is burned in the pulverized form.) However, this method is not very effective. Since the coal is so highly pulverized, there are a tremendous number of coal particles present. Because of the very small size, the coal is burned in a few seconds and the products are swept out of the furnace. Even though several times the theoretically required amount of lime or limestone is used, it is not possible to bring a sufficient amount of it in contact with the burning coal to remove more than 50 percent of the sulfur. Furthermore, the problem of cleaning the gases formed by combustion becomes a very serious one.

The turbulent bed method shows more promise. In this method, crushed coal and limestone are added to a bed which is agitated by the combustion air.

Some work is being done in investigating a process based on dissolving the coal in molten iron to which lime or limestone has been added. Sufficient air is supplied to produce carbon monoxide which is then burned in the combustion space of the furnace. Neither of these two methods has been tried in full sized plants.

Today the major effort towards sulfur dioxide control is being directed to its removal from the flue gas. Here the common approach has been to scrub the gases by passing them through a water slurry containing limestone, lime, dolomite, or other chemicals such as sodium compounds. Some work has been done both here and abroad to use char to absorb the sulfur dioxide. In several of these methods, attempts are being made to recover the sulfur, either in the elementary form or as sulfuric acid. In general, the economic desirability of recovering the sulfur has not at present been demonstrated.

Many problems have been encountered in flue gas desulfurization. First of all, the volume of gases to be handled is tremendous. For example, when 250 tons of coal per hour are burned, the volume of the gases to be handled may be approximately 35,000 cubic feet per second. Frequently, mechanical difficulties, such as scaling and corrosion, are encountered. Disposal of the vast amount of residue will be a major problem, particularly for plants located in built up areas. The addition of sulfur dioxide removal equipment may increase the initial cost of a power plant by as much as 20 to 25 percent. The cost of operating this equipment will materially increase the cost of producing power. Since sulfur dioxide removal equipment has not been operated extensively in full size power plants, both the initial and operating costs now being quoted should be taken as a very rough estimate at best.

In addition to the problems of mining coal and burning it, two other problems must be considered. At present we do not have the trained miners available to materially increase our coal output, particularly from underground mines. Unless working conditions and miners' wages are improved, it will be difficult to obtain sufficient capable manpower to greatly increase our production of coal.

Three to five years are required to open up a new large mine. Vast sums of money are needed to sustain operations before coal can be sold. Manufacturing facilities must be increased to produce the mining equipment required for increasing the output. In addition, the rolling stock of the railroads must be greatly expanded to transport the coal. All of this means that vast amounts of money and resources must become available if we are to greatly increase our output of coal.

6.5 PEAT

As discussed earlier, coal was developed by nature from peat. Although peat thus is not coal, it will be discussed here since it can, after being dried, be burned in a manner similar to coal and has some potential for replacing coal.

It is very difficult to estimate the total amount of peat in the United States. Based on the best available estimates it has been stated that peat is our second most abundant fossil fuel (after coal). Much peat can be found in the northern plain states and also in the mid-Atlantic coastal region, particularly North Carolina. In the past, it has not been used as a fuel in the Western Hemisphere. However, it is an important energy source in such countries as Ireland, Finland and the Soviet Union.

In 1976 the Institute of Gas Technology, under the sponsorship of the Department of Energy, began work to investigate the feasibility of producing synthetic natural gas by gasification of peat. Studies are being made in Minnesota of the characteristics of peat and the feasibility of mining it. The Grand Forks, North Dakota, Department of Energy Laboratory has begun funding to investigate the feasibility of burning a mixture of peat and lignite coal. Some effort also is being made in Maine to use peat.

Efforts have been initiated in North Carolina to try to use the large deposits of peat in this state. It has been estimated that North Carolina contains more than 800 million tons (moisture free) of peat on some 860 thousand acres of land. Theoretically, this peat could supply all of the state's total energy needs for more than ten years. Peat is now being mined at the rate of about 2,500 tons per week. There are plans to use peat in one or more steam power plants. Other possible uses are as a fuel in firing brick kilns (North Carolina is the largest producer of bricks) and as a source of synthetic natural gas and possibly methanol to blend with gasoline for automotive use.

There are certain problems involved in the use of peat. In general, the moisture content of peat as surface-mined is very high. This difficulty can be partially overcome by depositing the peat on the surface as it is mined and allowing it to be sun dried. However, much of the peat in North Carolina is found at very low elevations where there is the possibility of water intrusion as the peat is mined. The rising costs of other fuels, though, has created an incentive to try to overcome these difficulties and to use peat extensively.

6.5 SUMMARY ON COAL

Our known reserves of coal greatly exceed those of oil and natural gas, and hence offer the potential of significantly relieving our energy crisis. Even if mined at greatly increasing rates, those reserves which can be mined economically with present technology are sufficient to last us for several centuries. In addition, we have other reserves whose use will become economical and which will last for several additional centuries if energy costs become much higher than at present. This will be particularly true if mining technology can be improved.

At present, approximately half of our coal is obtained by strip mining. Much effort and money must be expended after strip mine operations are completed at a given site to restore the land to acceptable conditions. In the western Great Plains and Rocky Mountain regions, where large quantities of lignite and subbituminous coals are acceptable for strip mining, a shortage of water makes it difficult to grow vegetation on restored land.

Underground mining is a hazardous occupation. Much effort and money must be expended to prevent fires, explosions and roof cave-ins as well as to protect miners from black lung disease. In addition, much effort and money must be expended to dispose of both mine refuse and mine water without seriously damaging the environment.

Unless great efforts are made to prevent it, excessive amounts of particulate matter and sulfur will be discharged into the atmosphere when the coal is burned. Nitrogen oxides are also formed which cause some harmful effects. All of these adverse effects can be minimized but the cost will be very high.

Approximately 80 to 85 percent of the coal presently burned in this country is burned east of the Mississippi River. Most of it is bituminous coal. Over 80 percent of the eastern coal contains too much sulfur to meet EPA standards unless methods are used to reduce the amount of sulfur dioxide emitted when this coal is burned. Many methods have been devised but all methods materially increase the cost of producing power and, furthermore, operating difficulties are encountered as well as problems in waste disposal. Although there are large supplies of low sulfur lignite and subbituminous coals located in the western part of the country, their low heating values and high ash and moisture content presently make it uneconomical to ship these coals to the east to supplant the high sulfur bituminous coals.

As indicated in Table 2-1, it is predicted that the use of coal will increase by approximately 100 percent in the next 10 to 15 years. However, vast sums of money will be required to open up new mines, to safeguard the health of the miners, to make new mining equipment, to construct coal hauling equipment, and to protect the environment. Careful planning will be required since it will take 3 to 5 years to open a new mine.

SUPPLEMENT

PROBLEMS

6-1. A two-unit coal fired plant burns 400 tons of coal per hour. Determine the number of freight cars of coal which must be delivered to the plant per week, assuming there are 60 tons of coal per car.

6-2. The coal in Problem 6-1 has a sulfur content of 2.5 percent. Determine the amount of sulfur dioxide discharged by the plant per week. A pound of sulfur forms two pounds of sulfur dioxide.

6-3. If the maximum allowable amount of sulfur dioxide to be formed per million Btu heating value of the coal is 1.2 pounds, determine the maximum allowable percentage of sulfur in a coal having a heating value of

a. 13,000 Btu per pound
b. 9,000 Btu per pound

6-4. The gases from a power plant burning 200 tons of coal per hour are to be passed through a slurry of water and limestone to reduce the amount of sulfur dioxide. Determine the volume of gases to be handled per second. There are 13 pounds of gases per pound of coal. The density of the gases is 0.054 pounds per cubic foot.

6-5. Many years ago, coal was used very extensively for home heating. Do you think that coal may be used again for this purpose sometime in the future? Why?

6-6. The coal in Problem 6-1 is 11 percent ash. Determine the number of tons of ash to be handled per week. Note: much of this ash must be removed from the gaseous products of combustion.

6-7. As shown in Table 2-1, we export a considerable amount of coal. Should we not stop all exports of coal now to provide energy for ourselves in the future?

6-8. A power company is located wholly in a very large city. Presently it is burning oil in all of its plants. List problems to be encountered if it converts to coal.

7

Oil*

7.1 OIL AND ITS PRODUCTS

Many theories have been advanced as to the origin of oil. The most plausible theories postulate that oil was of either vegetal or animal origin. The vegetal theory holds that there was decomposition of vast quantities of gelatinous algae or marine vegetation. Upon decomposition, a gas, largely methane, and an oil were formed. If the formation occurred in sandy porous soil, the oil was carried away from place of formation by gravity or by water. The animal theory postulates that the oil was formed by the decay of vast quantities of marine animals, such as fish, mollusks and corals.

Oil, or petroleum, is widely distributed throughout many areas. The prerequisites for finding oil in commercial quantities include: (1) porous rock to contain the oil, (2) impervious rock to prevent its escape, and (3) a geological structure to permit the oil to collect in the given location after it was formed elsewhere. A schematic sketch of an oil and gas field is shown in Figure 7-1.

Crude oil is composed of very many different types of hydrocarbon molecules. The relative amount of any given compound present in a particular crude is dependent on the source of the crude. Partial separation of the various types of molecules may be done by fractional distillation. Here, the crude oil is heated, and the more volatile compounds such as naphtha and gasoline, being driven off (i.e., vaporized) first. As the heating is continued, the heavier† fractions, such as kerosene, number 2 fuel oil, diesel fuel, and then the heavier fuel oils are driven off. The portion of the crude which is left after the heating process is known by various names, such as residual oil, bunker C oil, or number 6 fuel oil. Its composition varies greatly, depending on the source of the crude and also on the extent of the heating process. It will contain dirt as well as tars and other asphaltic compounds.

---

* Following common practice, the term "oil" as used here refers to petroleum products only but does include natural gas liquids.

† Heavy refers to the high density, high boiling fractions.

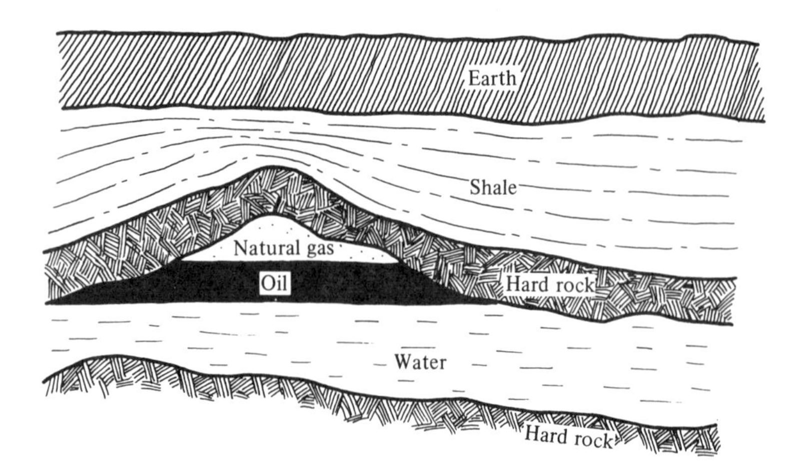

Figure 7-1. Oil-gas field

Because of the much larger demands for the lighter fractions, particularly gasoline, normally some of the heavier compounds are catalytically cracked. This is accomplished by heating the oils in the presence of a catalyst. Although theoretically it is possible to produce almost any distribution between light and heavy products by catalytical cracking, once a refinery is built the percentage of a crude which can be turned into a given product, such as gasoline, is limited. The refiner does have some control on the properties of a given product by using different crudes and by blending.

7.2 THE IMPORTANCE OF OIL

Oil, including its various products, is an excellent fuel. Because it is a liquid it can be transported readily. It can be burned easily in household and other types of furnaces although its efficiency normally is lower than that of natural gas. On the other hand, except for very large furnaces, oil fired furnaces are more efficient than coal fired ones. Oil is cleaner and more easily handled than coal. Because of these factors, there was a shift from coal to oil and gas in residential furnaces several decades ago. More recently, there has been a shift in industrial and commercial installations from coal to oil and gas.

Much of the oil produced in this country has a sufficiently low sulfur content to meet EPA sulfur dioxide regulations. The remainder of the oil, some 20 percent of the total, can have its sulfur content reduced to acceptable values by chemical treatment. In spite of added cost of the treated oil, many large industrial users and power plant operators switched from coal to residual oil. In some cases, this switch was made mandatory by state or local government action to control sulfur dioxide emissions.

We rely most heavily on oil for our total energy needs. As seen in Table 2-1 oil supplied 47 percent of the energy which we used in 1979. In that year oil supplied substantially all of the energy required for transportation, 28 percent of the energy required for residential use, 38 percent for commercial use, and 35 percent for industrial use. Approximately 15 percent of the energy required for electric power generation came from oil. Oil supplies the major portion of our needs for such non-energy uses as the petrochemical industries (plastics, tires, etc.).

In 1979 we imported approximately 47 percent of the oil which we used. The amount of oil imported in 1980 was substantially less. However, due to the great increase in the cost per barrel, we paid more total for our imported oil in 1980 than we did in 1979. Although it is anticipated that we will decrease, somewhat, our demand for foreign oil in the next decade or two, the great increase in the cost of this oil will mean that we will pay much more than 100 billion dollars a year for it—about 425 dollars for every man, woman and child in this country. Importing this oil has a serious adverse effect on our economy. Moreover, if this oil were to be cut off, as it was in the fall of 1973 by the Arab oil embargo, another serious recession would take place.

7.3 OUR OIL RESERVES

Because oil is presently so essential to our way of life and undoubtedly will continue to be for years to come, it is necessary to know the magnitude of our oil reserves.

As a basis for comparison, we used approximately 6.38 billion barrels of oil in 1979, of which about 47 percent was imported.

Table 7-1 shows that years ago we were finding new supplies faster than we were using the oil. Since 1970, however, in spite of extensive drilling operations, we have been taking oil from the ground faster than we can find new sources.

Table 7.1 Proved Reserves of Crude Oil (Billions of Barrels)

	Crude Oil	Natural Gas Liquid	Total
1950	25.3	3.0	28.3
1960	31.6	4.7	36.3
1970	39.0	5.3	44.3
1978	27.8	4.1	31.9

Note: The amount of natural gas liquid is reported in terms of the amount of crude oil equivalent.

Based on *Reserves of Crude Oil, Natural Gas Liquids and Natural Gas Liquids and Natural Gas in the United States and Canada and United States Productive Capacity as of December 31, 1978.* Published jointly by American Gas Association, American Petroleum Institute, and Canadian Petroleum Association, Volume 33, June 1979, pages 22-24, 110 and 126.

Table 7-2 lists the estimated undiscovered but recoverable oil resources. It should be emphasized that these figures are estimates at best and subject to

Table 7.2 Estimated Undiscovered Recoverable Crude Oil Resources (Billions of Barrels)

Onshore	
Continental U.S.	31-63
Alaska	6-18
Offshore	
Pacific Coast	2-5
Gulf of Mexico	3-8
Atlantic Coast	2-4
Alaska	3-31
Grand Total	50-127

Note: These figures do not include a total of 11 to 22 billion barrels of natural gas liquids. The totals were obtained by statistical methods rather than straight additions.

Based on *Geological Estimates of Undiscovered Oil and Natural Gas Resources in the United States*. Circular 125, U.S. Department of Interior, U.S. Geological Survey, Washington, D.C., 1975.

considerable revision. Some years ago, the U.S. Geological Survey estimated that there were vast quantities of oil, perhaps some 200 to 400 billion barrels, recoverable and nonrecoverable, offshore in the continental shelf of the eastern United States. Later, the figure was greatly reduced. (See Table 7-2.)

The total amount of oil recoverable from any oil field is dependent on the nature of the particular field. In general, however, only about a third of the amount of oil in a given field is recovered by present technology. It is possible to increase the yield from a given field by almost a third of that normally produced. Several methods are available. Hydraulic fracturing may be used to break up pockets of oil and permit it to drain to the bottom of the well. This fracturing is accomplished by injecting high pressure water (perhaps at several thousand pounds per square inch) into the well. Alternatively some fields lend themselves to flooding. In this method water is introduced into the well. Since the oil floats on water, it will be floated out of low spots, pockets and crevices in the rocks. Some crude oil is so viscous that much of it will not drain towards the bottom of a well. In such cases, steam may be introduced into the rock structure surrounding the well, thus heating the oil and reducing its viscosity sufficiently so it will flow. It is possible to introduce certain chemicals such as detergents to absorb the oil.

Some of these methods have been used to increase the output of certain wells. Up to recently, however, the added cost of obtaining additional amounts of oil from given wells often exceeded the price which could be obtained for it. Now that the control price of domestic crude has been lifted, it is expected that it will become economically feasible to reclaim much more oil from a given well.

In addition to increasing yields of existing wells, new wells continue to be drilled. In 1979, 19,330 successful oil wells were drilled compared with 13,020 in 1970. The average depth of oil wells drilled in 1979 was 4,054 feet and cost an average of $208,000. At present, at least 13 percent of our domestic oil is obtained offshore. This percentage is expected to increase. The cost of drilling offshore wells is many times that of onshore wells. The offshore wells, in general, are much deeper—at least 10,000 feet. Offshore wells are expected to become even deeper, making drilling costs much more expensive. In considering costs of finding new oil sources, consideration must be given to the costs of drilling dry wells. In 1978 the total cost of the dry holes, both oil and gas, was 4.43 billion dollars.

It should be evident that it is costly both to increase the yield of old wells and to search for oil in new wells. Although oil companies must not be permitted to make excessive profits, they cannot be expected to try to produce more oil at prices below the cost to them. We can continue to drift along, importing more and more oil at higher prices. However, it is preferable that the government examine the total situation and then make decisions as to how more oil could be produced if oil prices were to rise without undue profits for oil companies. The government must resist pressures from oil companies for uncontrolled, unreasonable prices. But it must also resist pressures from the general public which demands unlimited amounts of oil at low prices. The situation is complicated by the fact that if oil production is increased now, less oil will be available in the future.

7.4 SUMMARY ON OIL

Oil is a versatile fuel. It can be refined into many different products, thus satisfying a wide range of needs. In 1979, oil supplied 47 percent of our energy demands. The various sectors of our economy, particularly transportation, are heavily dependent on the availability of oil. It is difficult to perceive any changes which would decrease our dependence on oil for the next decade or two. However, our domestic oil resources are limited. From the information presented in this chapter, it can be seen that we may encounter drastic shortages of domestic supplies in the next two or three decades. If this be the case, we may find it necessary to import over 100 billion dollars worth of oil by 1990. It is very difficult to envision how it will be possible to pay for this amount of oil without drastically upsetting our economy. Although it is possible to increase our own production of oil temporarily, we must be prepared to pay higher prices for this oil.

In short, this country should face up to the fact that it will become even more costly, perhaps drastically so, to continue our ever growing demands for oil, the energy cornerstone of our economy. We must try to find some way to decrease our dependence on oil two or three decades from now.

SUPPLEMENT

PROBLEMS

7-1. Oil is being considered for the power plant in Problem 6-1. The oil has a heating value of 142,000 Btu per gallon. There are 42 gallons per barrel. Determine the number of barrels of oil required per week. Assume that the efficiency of the two plants are equal and that the coal has a heating value of 13,000 Btu per pound.

7-2. If the oil in Problem 7-1 costs 85 cents per gallon, determine the weekly cost of oil.

7-3. The coal in Problem 6-1 costs $40 per ton. Determine the weekly coal cost and compare with Problem 7-2.

7-4. Explain the statement in Section 7.2 about the comparative efficiencies of oil and coal fired furnaces.

7-5. Why do you feel that only about one-third of the oil in a given field is now recovered?

7-6. If our population in the year 1985 is 250 million and we should import 8.5 million barrels of oil per day at that time at a cost of $35 per barrel, determine the cost to each man, woman and child in this country per year to pay for this oil.

7-7. a. It requires 985 gallons of oil per year (140,000 Btu per gallon) to heat a house. The oil costs $1.20 per gallon. Determine the yearly costs.

b. It is suggested that direct electric heating be used in place of the oil in part a. Assume that the efficiency of the oil furnace is 70 percent. Electric power costs 6 cents per kwh. Determine the yearly cost for direct electric heating.

7-8. It is suggested that a heat pump (see Appendix for a discussion of the heat pump) be used to heat the house in Problem 7-7. The heat pump delivers 2.5 times as much heat as the equivalent of the electrical input. Determine the yearly cost of heating with a heat pump. Note: In very cold weather, it may require electric strip heating to augment the heat pump, thus increasing the total costs of supplying heat.

8

Natural Gas

8.1 NATURAL GAS USE

As with oil, various theories have been advanced as to the origin of natural gas. It seems reasonable to believe that natural gas was formed simultaneously with oil by the decay of either vegetal matter or animal matter.

Natural gas is largely methane, CH_4, although some natural gases may contain sizeable amounts of such light hydrocarbon vapors as ethane, C_2H_6. Its heating value is roughly 1,000 Btu per cubic foot under standard conditions (14.696 psia, 77°F).

In many ways, natural gas is an ideal fuel. Being gaseous, it does not discharge particulate matter. In general, the sulfur content is sufficiently low that it does not form excessive amounts of sulfur dioxide. When mixed with the right amount of air for combustion, it is relatively easy to prevent the discharge of significant amounts of carbon monoxide and hydrocarbons.

Originally, natural gas was considered a waste, to be disposed of when oil was produced. It was common practice to flare off the gas given off from oil wells. When its potential value was recognized, natural gas pipelines were constructed, now crossing most parts of the country. With the advent of these pipelines, natural gas almost completely supplanted manufactured gas. Because of its very desirable properties, the use of natural gas has greatly expanded in recent years. As shown in Table 2-1, we used almost 1¾ times as much natural gas in 1979 as we used in 1960. Although natural gas now supplies 25 percent of our energy needs, it is anticipated that this figure will drop to about 15 percent in 1995.

As may be seen from Table 8-1, approximately 40 percent of our natural gas is used for industrial purposes, 39 percent for residential and commercial purposes and 18 percent for electric power generation. In addition to being such a good fuel, natural gas is, in general, the cheapest source of energy except in those states in which it is produced. The reason for this is that the price of

Table 8-1. End Use of Natural Gas (Trillion Cubic Feet Per Year)

	1960	1965	1970	1975	1979
Residential	3.10	3.90	4.84	4.92	5.00
Commercial	1.02	1.44	2.40	2.51	2.65
Industrial	5.77	7.11	9.25	8.36	7.83
Electric Utilities	1.72	2.32	3.93	3.16	3.49
Transportation*	0.35	0.50	0.72	0.58	0.52
Total	11.97	15.28	21.14	19.54	19.49

* Primarily pipe line fuel.

Compiled from various sources and reported in *Annual Report to Congress, Volume 2,* Energy Information Administration, 1979.

natural gas has been carefully controlled at a rather low value for interstate use but there are no such price controls presently on intrastate gas.

As can be seen from Table 2-3, natural gas now supplies about 15 percent of the energy input for electric power generation. Much of this use occurred in the gas producing states where the uncontrolled prices induced the gas producers to supply large quantities of gas to the power companies. The use of natural gas for electric power production is to be questioned when the demand for it for other purposes is so great.

Although some industries can switch to another fuel (presently at a much higher price), other industries find it difficult to do so. Many fertilizer producers depend on natural gas not only for energy but also for its hydrogen needed for fertilizer production. Some industries, particularly textiles, depend on natural gas as a source of clean, non-polluting heat not available from other fossil fuels unless very major and costly changes are made in the methods of operation.

In general, industrial gas users have had interruptible contracts, which were desirable when gas supplies were ample since they were able to buy gas at very low rates. Now, however, the interruptible clauses mean that the industrial user is the first to be cut off in case of short supplies. In recent years, the demand for natural gas has greatly exceeded the supply, particularly during the winter months. The gas producers have chosen to supply their intrastate customers, since these customers pay what the producers feel is a reasonable price. As a result, the industrial users in the non-gas-producing states have been hurt by the lack of gas. Unless it becomes possible to increase the total supply of natural gas or to divert some of it from electric power production, many industries, particularly those which find it very difficult to convert to other fuels, may be forced to cease operations in the winter. For those states whose industries depend heavily on natural gas, this situation can be very serious, particularly in winters colder than normal.

Undoubtedly, if the price of natural gas were to be decontrolled, the gas producers would feel the incentive to search for new sources of gas and to drill more wells. Thus, it would be possible to halt the decline in gas production, and there might even be an increase in the output. As will be discussed next, the apparent supply of natural gas is quite limited. Hence, if the output of natural gas were to be increased now, this would only hasten the time when our natural gas supplies become depleted.

Some arguments have been made that there are ample supplies of natural gas in the ground but the gas is not being made available to the public because of the greed of producers. However, as can be seen from Table 8-2, we are using natural gas at a much faster rate than we are discovering new supplies. Almost 15,000 gas wells were drilled in 1979 compared with about 4,000 in 1970. (Note from Table 8-1 we used 19.49 trillion cubic feet of natural gas in 1979).

Table 8-2. Proved Reserves of Natural Gas (Trillion Cubic Feet)

1950	184.6
1960	262.3
1970	290.7
1978	200.3

Based on *Reserves of Crude Oil, Natural Gas Liquids and Natural Gas in United States and Canada,* American Gas Association, American Petroleum Institute, and Canadian Petroleum Institute, 1979.

8.2 NATURAL GAS RESOURCES

If we continue to use natural gas at the present rate, we will exhaust our proved reserves in 10 years. It may appear, however, from Table 8-3, using the high figure of 655 trillion cubic feet, that we will have ample supplies of natural gas for many years. This is far from the realm of probability since it is stated that there is only one chance in twenty that this amount of natural gas exists.

Table 8-3. Estimated Undiscovered Recoverable Natural Gas Resources (Billions of Barrels)

Onshore	
Continental U.S.	248-449
Alaska	16-57
Offshore	
Pacific Coast	2-6
Gulf of Mexico	18-91
Atlantic Coast	5-14
Alaska	8-80
Grand Total	322-655

Note: The totals were obtained by statistical methods rather than straight addition.

Based on *Geological Estimates of Undiscovered Oil and Natural Gas Resources in the United States,* U.S. Department of Interior, U.S. Geological Survey, Washington, D.C., 1975.

In 1977 we received almost one trillion cubic feet of natural gas from Canada. It is expected that we will continue to receive gas from Canada for some time but at a reduced rate. With the discoveries of sizeable amounts of natural gas in Mexico, plans are being made to pipe some of this gas to the United States. In some oil producing countries, natural gas is still treated as a waste product when accompanying the oil that is pumped from the ground. This natural gas is flared off, much as was done in this country many years ago. Now, growing attempts are being made to recover the natural gas. It is cooled down to at least −250°F and subjected to high pressure, thus liquifying it. As a liquid, it can be transported overseas in specially designed ships. In 1977 Algeria and Libya shipped about one-third of a trillion cubic feet of natural gas in this manner.

Many years ago, before the advent of natural gas pipelines throughout this country, large quantities of gas were produced from coal. Now, with the growing shortages of natural gas, a considerable effort is being devoted to produce appreciable amounts of gas from coal. The amount of gas which will be produced from coal in the coming years is very difficult to predict. The subject of the production of gas from coal will be discussed in Chapter 20.

8.3 SUMMARY ON NATURAL GAS

Natural gas in many ways is an ideal fuel. It is easy to burn and does not readily pollute. Since its cost has been controlled for interstate use, in general it is by far the cheapest source of energy. As a result, the demand for natural gas in the last quarter of a century has increased several fold.

Although much natural gas is known to remain underground and more gas probably will be discovered, all indications are that the amount of natural gas produced in the continental United States will continue to decrease. If the price of natural gas were to be decontrolled, its production rate should increase, temporarily. Such action, however, would only decrease the amount of natural gas available for future use.

Three factors should help alleviate the natural gas shortage in the next decade or two:

1. It is projected that a sizeable amount of natural gas will become available from northern Alaska by the mid 1980's.
2. It is anticipated that appreciable amounts of liquefied natural gas will be imported in the next decade or two.
3. Pilot plant demonstrations indicate that significant amounts of gas will be produced from coal in the next two or three decades.

In spite of these three developments, it appears that gaseous fuel will be in very short supply by the start of the twenty-first century. If this be true, careful planning should be made to use our present supplies wisely.

SUPPLEMENT

PROBLEMS

8-1. Natural gas is being considered for the power plant in Problem 6-1. The gas has a heating value of 1,020 Btu per cubic foot. Determine the number of million cubic feet required per week. Assume that the efficiencies of the two plants are equal.

8-2. If the natural gas in Problem 8-1 costs $1.75 per thousand cubic feet, determine the weekly costs of the natural gas. Compare the results with those of Problems 7-2 and 7-3.

8-3. A family uses 80 gallons of hot water per day. The temperature of the water is increased 70°F as it is heated. Natural gas having a heating value of 1,020 Btu per cubic foot is used for heating. It is burned with an efficiency of 75 percent. It costs $1.75 per thousand cubic feet. Calculate the cost of gas per month (30 days).

8-4. Same as Problem 8-3 except electrical energy is used for heating. The efficiency is 95 percent. The cost of electricity is 6 cents per kWh. Compare your answer with that of Problem 8-3.

8-5. It is proposed that the use of natural gas for electric power generation be prohibited as soon as it is reasonably possible to convert present natural gas fired power plants. Should this be done? Why?

8-6. Should the price of natural gas be deregulated? Why?

8-7. Presently (1982), there appears to be a surplus of natural gas. Do you recommend that users of other means of heating should convert to natural gas?

9

Nuclear Power

9.1 INTRODUCTION

Nuclear power has become quite controversial. On one hand, it has been stated that nuclear power can make a very major contribution towards solving our future energy problems. On the other hand there are those who state that nuclear power is so hazardous that it should be completely abolished. Unfortunately, many of those who have strong opinions on the subject have made little effort to understand the total aspects of nuclear power.

It is the purpose of this chapter to examine the mechanism of nuclear power production, the potential of nuclear power as well as the hazards associated with it.

No conclusion should be drawn as to the desirability of using nuclear power until a careful study is made of the potential of other energy sources for meeting our future energy needs. This is particularly true for alternative energy sources, which will be discussed later. Can these alternative energy sources be developed in the next one or two decades to meet many of our energy needs at a reasonable cost? If the answer to this question is a positive yes and not just a hope, it will be very difficult to justify the expansion of our use of nuclear energy. If the answer is no, how can we obtain a sufficient amount of energy for our future needs without nuclear power?

9.2 NUCLEAR FISSION

Thus far we have been concerned with energy released by combustion of fossil fuels. This action is a chemical one in which the atoms of fuels react with atoms of oxygen, generally from air, to form new compounds. The energy released by this action is known as chemical energy.

We will now consider a different source, nuclear energy. This energy is obtained by splitting atoms of a substance. In this process there is conversion of mass into energy as given by the Einstein equation

$$e = mc^2 \tag{9-1}$$

where e is the energy released, m is the mass converted to energy, and c is the velocity of light. When m is expressed in grams and c is 2.9979×10^{10} cm per sec, e will be in ergs. The process of splitting atoms is called *fission* and the material which can be split is said to be *fissile*. The fission process involves atoms of very high molecular weight. This chapter will deal with this type of nuclear energy.

A second process of developing nuclear energy is that of *fusion*. In this process, atoms of low molecular weight substances are fused together. In the fusion process, there is also a transformation of mass into energy, the transformation being also governed by Equation 9-1. To date, the fusion process has been carried out in the laboratory only. It is felt by many that fusion will not become a commercial reality before the year 2000. Hence the discussion of fusion will be deferred until Part III, Our Alternative Energy Sources, Chapter 19.

The heart of the atom is the nucleus. The nucleus is composed of protons and, with the exception of hydrogen, of neutrons. The protons are positively charged. Surrounding the nucleus are electrons, which are negatively charged. The electrons move in fixed spherical orbits, called shells, around the nucleus. In the normal atom, the number of protons and the number of electrons are equal and hence the atom is electrically neutral. See Figure 9-1.

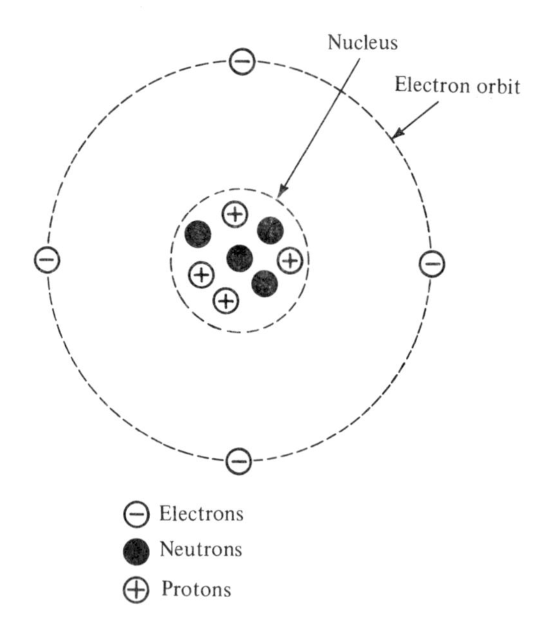

Figure 9-1. Structure of simple atom

A given element may have several *isotopes*. The various isotopes of a given element have the same number of electrons and the same number of protons but different number of neutrons in their nuclei. The chemical behavior of the various isotopes of a given element is the same since this behavior is a function of the electron structure. The atomic number of an element is the number of protons (and hence the number of electrons for a neutral atom) possessed by an atom. The mass of an atom is specified in terms of atomic mass units (amu). The atomic mass unit is established by taking the mass of the atom of the common isotope of oxygen as 16 amu.* The mass of the electron, then, is 0.00055 amu, the proton 1.00785 amu and the neutron is 1.0089 amu. The mass of a given atom is the rounded off sum of the mass of its electrons, protons and neutrons.

Of all the elements found in nature, only uranium is readily fissionable. Natural uranium is 0.7 percent ${}_{92}U^{235}$. This means that the atomic number is 92 and the atomic mass is approximately 235. By far, most of natural uranium is ${}_{92}U^{238}$, being 99.28 percent of the total. There is a trace of ${}_{92}U^{234}$. When dealing with a given element, it is common practice to omit its atomic number. Thus we say that natural uranium contains three isotopes, U^{234}, U^{235}, and U^{238}.

Only U^{235} is fissile and hence can be used to produce energy. U^{238} is *fertile* (i.e., it can capture a neutron. After a short period of time, it decays into plutonium, which is fissile).

U^{235} will fission rather readily when struck by low velocity neutrons, producing many lighter elements and also a large amount of energy. As it is fissioned, each atom of U^{235} releases on the average 2½ neutrons. The amount of energy released by the fission of one atom of U^{235} is approximately 200 million electron volts (200 meV). One eV (electron volt) is the energy acquired by an electron when it passes through a potential difference of one volt. One meV is equivalent to 1.18162×10^{-13} ft lb. This may seem to be an exceedingly small amount of energy. It is, but, since there are approximately 2.5×10^{21} atoms per gram of U^{235}, the complete fission of one gram of U^{235} will produce a heating effect of approximately 7.59×10^{7} Btu. A nuclear power plant consuming completely one kilogram of U^{235} will deliver approximately the same amount of electrical energy as will a fossil fuel plant burning 2,400 tons of coal.

As stated above, the U^{238} atoms readily capture the high velocity neutrons but only slow velocity neutrons will fission the U^{235} atoms. Thus it is necessary to slow down the neutrons as soon as possible after they are released. This is accomplished by introducing a moderator into the reactor. Various substances, such as water, beryllium, beryllium oxide, and graphite are satisfactory moderator materials.

Four things can happen to the neutrons which are freed when nuclear fission occurs:

1. They may be captured by the U^{238}.
2. They may be captured by the rest of the reactor.
3. They may escape from the reactor.
4. They may fission more atoms.

---

* In chemical work, the mass of the average oxygen atom is taken as 16 units.

In order that the fission process can continue, it is essential that one of the neutrons released upon each fission be available for further fission. The amount of neutrons which are produced is a function of the volume of the reactor. The number which will escape is a function of the surface area. The ratio of the surface area to volume decreases as the size of the reactor increases. Below a certain size, this ratio is so large that such a large percentage of the released neutrons escape from the reactor that fission cannot continue. When the reactor is sufficiently large that fission can continue, the reactor is said to be *critical.*

Control rods are an essential part of the reactor. The control rods may be made of cadmium, boron steel, or some rare earth, such as hafnium. The control rods readily absorb neutrons. When fully inserted in the reactor they absorb so many neutrons that fission cannot continue. The rate at which fission does occur is controlled by the extent to which the control rods are withdrawn from the reactor.

9.3 NUCLEAR FUEL PREPARATION

Uranium is a metal, just as lead, gold and iron are metals. Since pure uranium readily reacts chemically, it occurs in nature only in compounds. Uranium is widely dispersed. It makes up about two to three parts per million (ppm) of the earth's crust. Much of this uranium is found in rocks. In general, 500,000 pounds of rock contain on the average one pound of uranium. Certain rocks, such as granite, contain four ppm of uranium. However, uranium is not sufficiently concentrated in these sources to make it recoverable with present technology. Uranium is found in a more concentrated form in the Rocky Mountain region of this country, primarily in three compounds: uranium oxide (uranite or pitchblende), potassium uranium vanadate (carnotite), and uranium hydrous silicate (coffinite). These compounds are found embedded in rock in various concentrations. At present, it is not economically feasible to mine these rocks unless there is at least three or more pounds of uranium per ton of rock.

Some of these rocks occur at the surface or just below it. In this case, the overburden of rocks and dirt may be removed and the uranium-bearing rocks then recovered. Most of the uranium-bearing rocks, however, exist at such a depth that underground mining must be used to reclaim them.

After mining, the rock is first crushed and then leached by percolating chemicals through it to dissolve the uranium. The uranium is then treated with chemicals in various processes to form uranium oxides, primarily U_3O_8 (yellow cake). Since the concentration of U^{235} is so low (0.7 percent) the U_3O_8 must be enriched (i.e., the concentration of U^{235} must be increased to perhaps 3 percent) before it is suitable for power production. To accomplish this, the uranium is converted to uranium hexafloride, UF_6, which is gaseous. At present the common method of enrichment is the gas diffusion process. Since the U^{235} atoms are somewhat lighter than the U^{238} they possess slightly higher velocities and hence will pass a little more readily through some semi-permeable membranes. When the UF_6 is put through several hundred membranes in series, the final concentration of the U^{235} in the uranium may be about 3 percent. Because so much power is required to compress the gas and keep it under the desired pressure, other methods, particularly the centrifuge process, are being considered for uranium enrichment.

After enrichment, the UF_6 is converted into uranium oxide, UO_2, a powder. This powder is pressed into small pellets. The pellets are placed inside tubes, generally made of zirconium to stand the high temperatures and to resist radiation damage. The assembly is now known as a rod. The rods are small in diameter, perhaps a half inch or so, but may be as much as 12 feet in length. The rods are assembled together in bundles, and the bundles are placed within the reactor to constitute its core. See Figure 9-2.

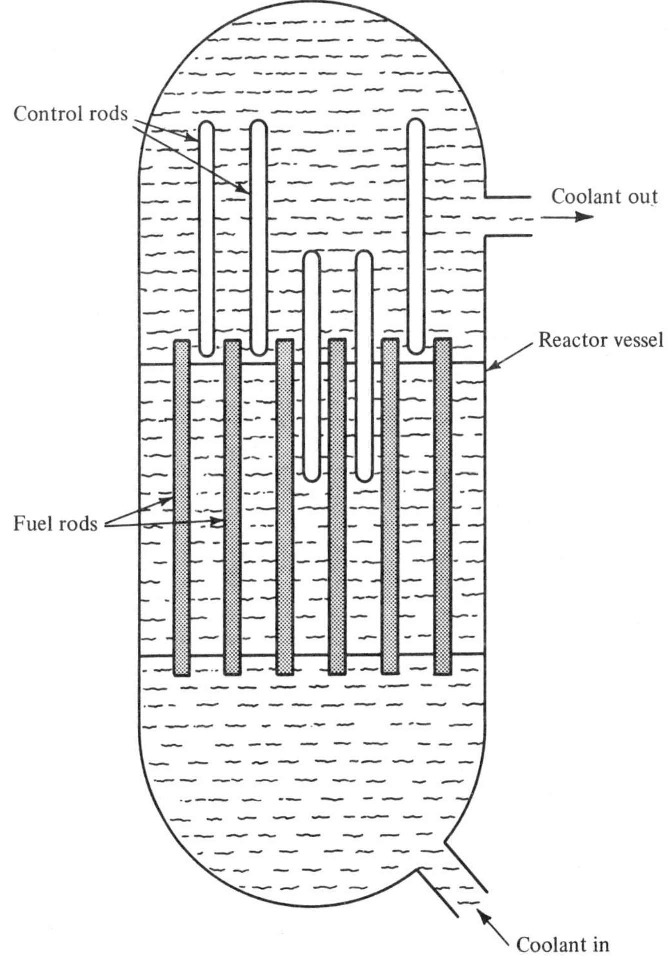

Figure 9-2. Fuel rods assembled in a reactor

9.4 TYPES OF REACTORS

It is possible to use various fluids to remove heat from the core of the reactor and, directly or indirectly, produce work and then electrical energy. In Europe, particularly in England, gas cooled reactors have been used extensively. Both carbon dioxide and helium can be used as the gas coolant. It is possible to pressurize the gas sufficiently so that it can be used in a gas turbine. However, the present allowable maximum temperatures for gas turbines limit the efficiency so much that this method is not being used now. Instead, the hot gases leaving the reactor are used to produce steam, which is piped to a steam turbine.

With one or two possible exceptions, all of the nuclear power plants in this country, at present, are light water reactors.* The light water reactors may be either the pressurized water or boiling water type. A pressurized water power plant is shown schematically in Figure 9-3. Water under pressures of perhaps 2,000 to 2,500 psia is pumped into the reactor and is heated close to its boiling temperature.† The high temperature water leaving the reactor flows through a steam generator. The steam generator is, in reality, a heat exchanger in which the reactor water produces moderate pressure steam as it is cooled down. The steam produced must be lower in temperature than that of the reactor water,

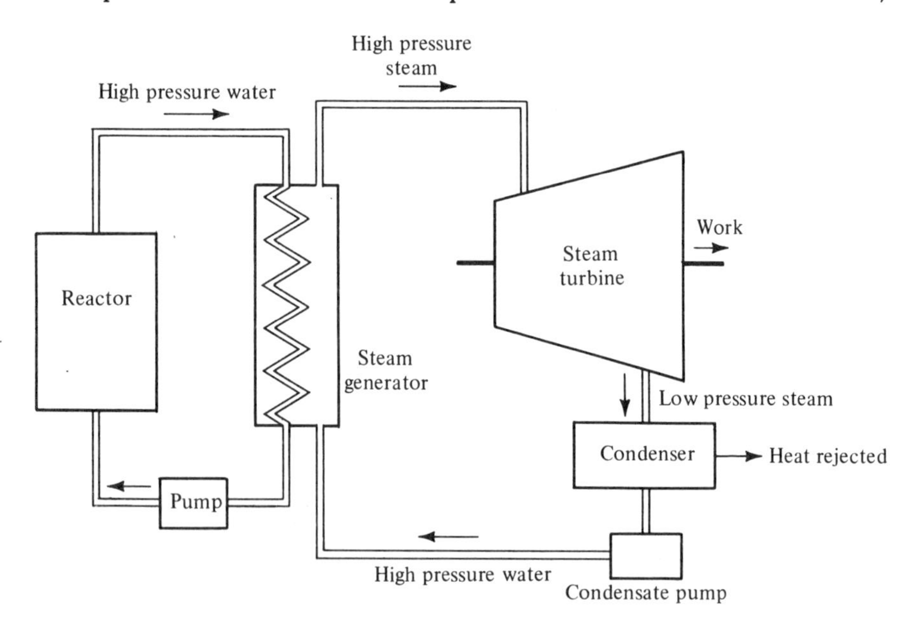

Figure 9-3. Elements of a pressurized water nuclear power plant

* Light water is ordinary water. The term is used to distinguish it from the heavy water reactors used to some extent in Canada. The heavy water molecules are made up of the heavy isotopes of hydrogen.

† At 2,000 psia, water boils at 636°F. At 2,500 psia, the boiling temperature is 668°F.

and hence its pressure must be lower than the water pressure in the reactor and is perhaps 1,000 to 1,200 psia. The steam from the steam generator passes to a steam turbine and thence to the condenser, as in an ordinary steam power plant.

The rate of water flow into a boiling water reactor is sufficiently low for the water to have time to turn into steam. Since the radioactivity of the steam is so very low, it can be used directly in the steam turbine. See Figure 9-4. The pressure of the steam is similar to that of the steam produced in the pressurized water reactor, being approximately 1,000 to 1,200 psia.

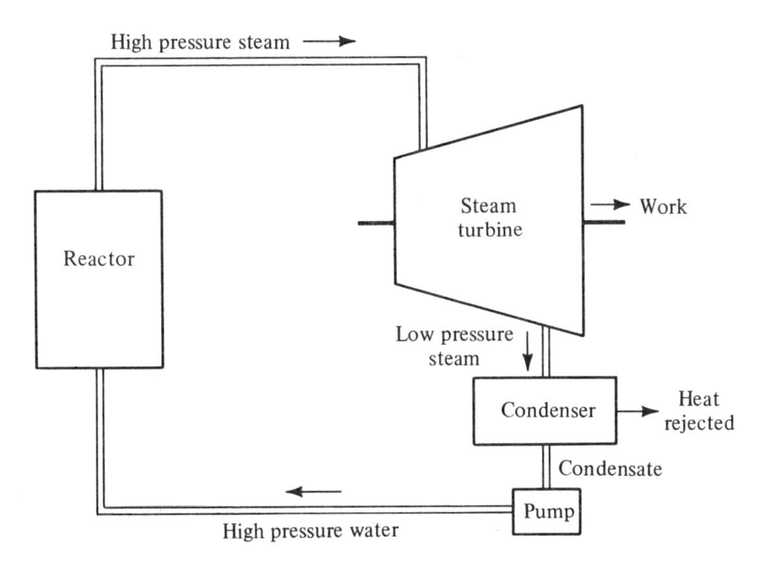

Figure 9-4. Elements of a boiling water nuclear power plant

Since the steam pressure entering the turbine of the light water reactor power plant is so much lower than that of a conventional fossil fuel power plant, the thermal efficiency of this plant is much lower, perhaps in the order of 32 percent compared with about 40 percent in the best of fossil fuel power plants. Not only must more fuel be used because of this lower efficiency, but also the problem of thermal pollution is much more severe. Some 40 to 50 percent more circulating water may be required for the condensers of a nuclear power plant than for a fossil fuel power plant having the same electrical output.

Elaborate precautions are taken both in the design and in operation of a nuclear power plant. Although there is considerable variation in the detailed design of various plants, in general, the reactor vessel is made of steel with walls perhaps one foot in thickness. In a moderate size plant the reactor vessel may have an outside diameter of 16 feet, a height of 40 feet and weigh, empty, a million pounds. It is common practice to enclose the reactor vessel in a containment shell, made up of several feet of reinforced concrete and also a steel casing. For the boiling water reactor, the steam generator may be placed within the containment shell. Figure 9-5 is a line diagram of the reactor portion of a nuclear power plant.

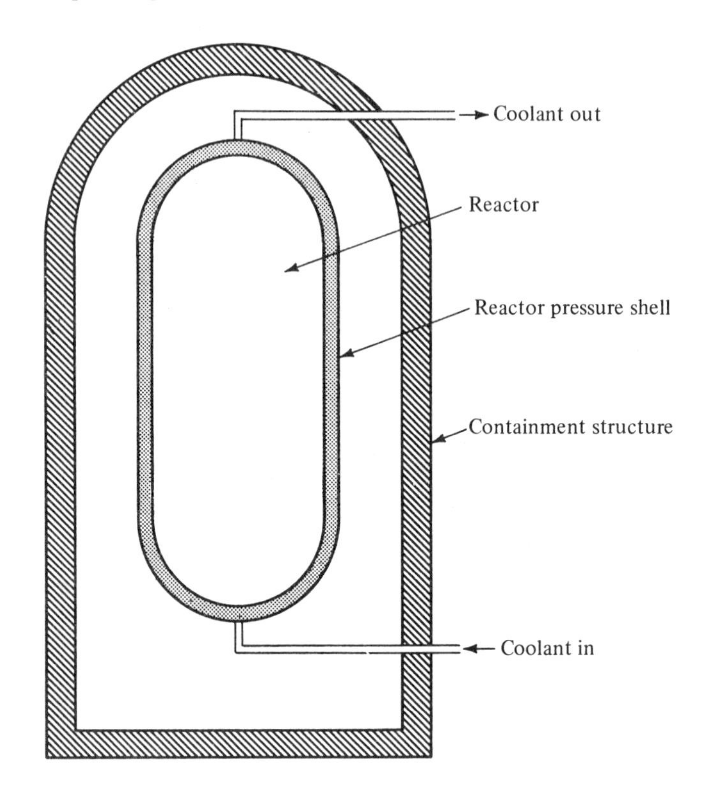

Figure 9-5. Nuclear reactor containment

Some effort is now being devoted in this country to developing a high temperature gas reactor (HTGR), using helium as the reactor coolant. The heated helium is used to produce steam for a steam turbine. Since the helium can be heated to a high temprature, it can produce steam at temperatures and pressures comparable with those in modern fossil fuel power plants. Because of the increased efficiencies of the HTGR plant, not only will less uranium be required but much less heat will be rejected in the condensers, thus helping to minimize the thermal pollution problem.

9.5 REPROCESSING AND WASTE DISPOSAL

The uranium present in new nuclear fuel rods is approximately three percent U^{235}. Operation of the reactor causes a decrease in the amount of U^{235}. When the concentration drops to about one percent of U^{235}, it becomes necessary to replace the fuel rods. Since the U^{235} content of the depleted rod is somewhat higher than that of natural uranium and since the depleted rod contains fissionable plutonium, it is desirable to reprocess it to obtain fissionable material. The half life* of some of the fission products in the depleted rod is very short but those of others are exceedingly long. To permit a decay of those materials having a short half life, it has been the practice to store the spent fuel rods under water for several months before reprocessing them. However, largely due to pressures of environmental groups, processing of the spent fuel rods has been halted, but it is anticipated that reprocessing will be resumed in the near future.

Several suggestions have been made for the disposal of the radioactive wastes remaining after separation in the reprocessing plants. Some of these suggestions involve burying them in deep holes in the earth, in polar ice caps, in the oceans, shooting them into outer space, and shooting them into the sun. Because of the various safety problems involved, none of these methods seems to be feasible. The two suggestions which are being given serious consideration are to store them in casks above ground or to place them underground in large salt deposits. Although radioactive wastes from nuclear weapon production have been stored as liquids in underground drums, it has been found that after many years, the drums may leak, thus permitting some of the radioactive wastes to seep into the ground. Although, apparently, none of this waste has as yet entered underground streams, the potential for this to occur does exist. For this reason, present plans are to store the liquid radioactive wastes under water for several years to permit a significant reduction in the radioactivity., Then the wastes would be converted to a solid and sealed in heavy casks for permanent disposal. Abandoned salt mines seem to offer an excellent place for disposal. The salt veins apparently have been very stable for thousands of years and salt will absorb the radiation readily and prevent its escape. It is also relatively easy to monitor the wastes. However, because of heavy pressures from various environmental groups, the government has deferred the use of salt mines.

Storage of the radioactive wastes above ground, although a real problem, does not appear to be as serious as some feel it is. Even if the use of nuclear power expands as rapidly as now contemplated, it has been stated that all wastes created by all plants in this country by the year 2000 can be stored on a 1,500 acre lot. The wastes would be stored as a solid in heavy steel casks. Provision would be made to remove the heat generated due to the decay of the radioactive materials. The casks would be guarded although, because of the heavy weight of the casks, only the largest of heavy duty material handling equipment could remove these casks from the depository. Some opponents of nuclear power feel that the waste depositories must be guarded for thousands of years, stating that plutonium has a half life of 24,600 years. However, plutonium is a very valuable fuel. Hence, for economic considerations alone,

* Half life is the length of time required for the radioactivity of a given substance to decrease to a half of its original value.

99.5 to 99.9 percent of the plutonium should be removed from the waste. When this is done, the radioactivity of the waste will decrease so much in a 600 year period that it will then be equal to that of natural uranium and hence cannot be particularly harmful.

It has also been suggested that there is a possibility of converting those elements having long half lives into ones with shorter half lives, thus reducing the potential danger from the radioactivity of the wastes. The feasibility of this proposal has not been demonstrated. There is the possibility that even if it becomes feasible to do this, an excessive amount of energy may be required to accomplish this conversion.

9.6 THE FAST BREEDER REACTOR

The breeder reactor is one which produces more fuel than it consumes. This is not, as it may seem, creating something out of nothing. Rather, it is designed to permit sufficient U^{238} atoms to capture high velocity (i.e., fast) neutrons (and then decay into plutonium) and thus to produce or breed more plutonium than the U^{235} which is used to keep the reactor operating. Theoretically, since U^{238} constitutes over 99 percent of the natural uranium the fast breeder reactor ultimately could produce over 100 times as much energy from a given amount of natural uranium as is now produced. Presently, however, the goal seems to be the production of about 50 times as much energy.

Not only does the fast breeder reactor offer the potential for producing much more power from a given amount of uranium but it can use much more expensive uranium without causing a significant increase in the cost of producing power. If the selling price of uranium were to be five to ten times the present price, it would become profitable to mine vast quantities of uranium ore that are not now economically desirable. It is for these reasons that much effort is being expended to develop the fast breeder reactor.

The fast breeder reactor differs significantly from present reactors. First, since the U^{238} atoms readily capture the high velocity neutrons but not the slow moving ones, the neutrons are not slowed down as they are in the present reactors. Second, reactor materials which are satisfactory for the low velocity neutrons may be completely unsatisfactory when used in the presence of high velocity neutrons.

European countries are far ahead of us in developing the fast breeder reactor. The French put an experimental fast breeder reactor of 20 MW capacity in operation in 1967. In 1973, the Phenix, a 250 MW fast breeder, was started up. The French are now building a 1,200 MW fast breeder, due for operation in 1982, and are planning to start construction of two 1,500 MW fast breeder reactors in the near future. The British have had a 250 MW fast breeder reactor in operation since 1974 and have plans for a 1,300 MW unit with construction to be started in 1985. Russia has had a 350 MW unit in operation since 1973 and a 600 MW unit was started up in 1980.

In this country, some years ago we started construction of a 330 MW fast breeder reactor on the Clinch River in the TVA region. Because, among other reasons, the fast breeder reactor creates large amounts of the highly radioactive plutonium, in 1977 President Carter halted the construction of this plant. However, Congress has continued to appropriate funds and electric power companies have also contributed money, allowing some work to continue on this plant. As with many completely new projects, cost over-runs have been high and many engineering problems have developed. Thus it is difficult to predict when this fast breeder will go into operation. There is some feeling that this reactor is becoming outdated and our efforts should be directed to newer fast breeder reactor plants.

Ordinary water commonly is used as the coolant in the conventional reactor. One reason for this is that it acts as a moderator to slow down the neutrons as they are released by fission. This is necessary since only the relatively slow neutrons will fission the U^{235} atoms. On the other hand, the neutrons must not be slowed down in the fast breeder reactor since the U^{238} atoms will capture only the fast neutrons. To avoid slowing down the neutrons in a fast breeder reactor the coolant used commonly is a liquid metal, commonly liquid sodium. This reactor is known as a liquid metal fast breeder reactor (LMFBR). Sodium creates problems since it is difficult to contain. It will explode if it comes in contact with water and may do so under some circumstances if it comes in contact with air. It does have an advantage over water in that it can be heated to much higher temperatures. This means that steam can be produced at temperatures and pressures comparable with those of fossil fuel power plants. Thus much higher efficiencies will be attained and much less heat is to be rejected in the condenser, thus minimizing the thermal pollution problem.

9.7 THORIUM AND THE SLOW BREEDER REACTOR

As has been stated, supplies of uranium for the conventional reactor are limited. The LMFBR has shown the possibilities of providing our needed electrical energy for centuries. However, this reactor creates large amounts of the highly reactive plutonium. Since, theoretically, it is possible that some of this plutonium could be high-jacked and used to make atomic bombs, some consideration is being given to the development of the LWBR (light water breeder reactor). The prime fuel for this reactor is thorium, Th^{232}. Thorium itself is not fissionable but fertile. It can capture slow moving neutrons, forming Th^{233}. This thorium will decay into U^{233} which is fissionable. The LWBR must be carefully designed. It is charged in its center with U^{233}. Two of the neutrons released by the fission of one atom of U^{233} must be used, one to continue the fission process and one to be captured by an atom of thorium to produce more U^{233}. After the original Th^{232} is nearly spent, the reactor fuel must be reprocessed to use the U^{233} which has been formed and to charge the reactor with more thorium.

Thorium is said to be relatively cheap and abundant, found in many foreign countries as well as in this country. In addition to the reserves of thorium in the western United States, thorium can be found in the sands of rivers and beaches in the eastern part of this country. It has been stated that some of the beach sand of South Carolina contains up to 9 percent of thorium oxide. It has been estimated that if we can recover a half of the energy of our known supplies of thorium, we will gain seven times as much energy as in the known world supplies of coal, oil, and natural gas.

At this time the economic feasibility of using thorium for power production is not known. In 1977, the old 60 MW Shippingport, Pennsylvania, conventional nuclear reactor was converted to a breeder reactor using thorium. It is anticipated that after some four years of operation, the reactor core will be removed and shipped to the Idaho National Engineering Laboratory for analysis. It is hoped that sufficient operating data can be obtained to show whether or not the LWBR is as good overall as or better than the LMFBR. It is possible that a somewhat larger thorium breeder reactor should be built and a LMFBR operator before a decision can be made whether this country should pursue a breeder reactor program. If so, the question will be which type.

9.8 URANIUM RESOURCES

As with fossil fuels, it is very difficult to make an accurate estimate of the extent of our uranium resources. Table 9-1 is taken from *Statistical Data of the Uranium Industry, January 1, 1979,* U.S. Department of Energy, Research Applications, Grand Junction, Colorado, pages 11 and 27. For many years, government action held the price of U_3O_8 to 8 dollars per pound. Present price is around 40 dollars per pound and rising. Note from Table 9-1 that it is economically feasible to recover much more uranium as the price increases. It has been estimated that about twice the amount of uranium can be obtained if its price was 100 dollars per pound instead of 50 dollars per pound. Under present conditions with the conventional reactor, if uranium cost is 100 dollars per pound, nuclear power will lose its competitive edge over coal produced electrical energy. However, the initial cost of uranium will have little effect on the cost of power produced in a fast breeder reactor power plant since perhaps 50 times as much power can be obtained from a given amount of uranium.

In 1979, 18.5 thousand tons of uranium were mined in this country. Of that amount, the net export was 0.9 thousand tons, leaving a net of 17.6 thousand tons. As predicted in Table 2-3, nuclear energy in 1995 will be between 4 and 5 times that of 1979. This means that unless the possible and speculative reserves become available, there will be a shortage of uranium in the next three or four decades. However, there will be a drastic change if the fast breeder reactor becomes a commercial reality in this country. Since the fast breeder can, ultimately, use 50 times as much of a given amount of natural uranium as can the ordinary reactor, with the fast breeder reactor we would have sufficient uranium to last for several centuries.

Table 9-1. U.S. Uranium Resources, January 1, 1979 (Thousands of Tons, U_3O_8

	$30 or less*	$50 or less*
Reserves†	690	920
Potential Resources	1,980	3,225
Probable	1,005	1,505
Possible	675	1,170
Speculative	300	550
Total	2,670	4,145

* The cost per pound of U_3O_8 does not include taxes, profit or amortization of existing capital equipment and hence is not the selling price.

† Does not include 140,000 tons of U_3O_8 estimated to be available as a by-product of copper and phosphate production, 1979-2000.

9.9 NUCLEAR REACTOR SAFETY

The uninformed public looks upon a nuclear reactor as a bomb which could explode at any time and cause as much destruction as an atomic bomb. For a nuclear plant located in a heavily populated area, they envision thousands of people killed and many more deformed for life. The truth is far from this. The present reactors are not bombs. They cannot explode as bombs do. Even for a freshly charged reactor, only about 3 percent of its uranium is fissionable. Neither can this material be made readily into a bomb. To do so would require almost as much effort as it would to start with natural uranium.

It has been argued that terrorists could organize a raid on an operating reactor. The outer containment shells of reactors, particularly those near airports, are designed to resist the impact of the largest loaded airplanes. Thus, ingress to the plant would have to be made by direct attack at the entrances. It should be rather easy to install such safety precautions which would shut down the plant completely in the event of this kind of an attack. The plant could be activated again only by an expert after the lapse of a considerable amount of time.

The question has been raised as to the possibilities of accidents occurring within the plant which could result in the total destruction of the plant and the spreading of radioactive material in the area around it. Such an accident could start with the loss of the reactant coolant, due to a break in the coolant pipeline. Although the control rods are designed to completely shut down all further fission automatically, there would be so much energy left in the reactor that, unless this energy were to be removed, the reactor would be melted and some of its contents might flow into the containing shell. It is conceivable that, in turn, some of the hot reactor material could melt through the containment vessel and thus flow to the outside and then release its radioactivity. To guard against the possible consequences of such an accident, backup devices are installed. For the light water reactor, a spray system may be used in which water is sprayed into the reactor vessel to cool it down. It is also possible to keep the space between the reactor vessel and the containment shell filled with ice. In case of this type of failure, the melting of the ice would provide sufficient cooling.

Such an accident did take place in one unit at the Three Mile Island Plant near Harrisburg, Pennsylvania, on March 28, 1978. Although some blame for this accident was placed on equipment, it appears that the accident was largely due to operator error in not leaving open the valve that controlled the water flow in the emergency cooling system and also not reacting as soon as possible when the emergency took place. Fortunately measures were taken to control the accident so that radiation did not spread beyond the plant itself. However, the reactor unit became so highly radioactive that it will take many years and millions of dollars to deactivate it.

To prevent such accidents in the future, the Nuclear Regulatory Agency has greatly tightened up operating procedures, particularly in requiring more intensive operator training. The power companies themselves are said to have set up industry-wide standards which exceed those of the Nuclear Regulatory Agency. With these added precautions, it is highly improbable that such an accident will happen again.

As with other types of power plants, equipment failures may occur outside of the reactor vessel. Such failures, although they may necessitate temporary shutdown of the plant do not affect the safety of the nuclear part of the plant. In spite of this, the news media may report such failures to the public, which then assumes that nuclear plants are so hazardous that their use should be prohibited.

Many questions have been raised about the probability of nuclear power plant accidents of such magnitude as to endanger the entire populace. In an attempt to answer these questions, the Atomic Energy Commission (AEC) initiated a two year program under the direction of Professor Norman C. Rasmussen of the nuclear engineering department at the Massachusetts Institute of Technology. The report, a very thorough one and costing about three million dollars, was delivered in August 1974. It contains 11 volumes. In spite of charges of critics that the report must be biased in favor of the AEC, there is no evidence to show that the AEC exerted pressure on the committee to come up with a favorable report.

Most of the accidents which may be conceived for nuclear plants are very improbable. For instance, Rasmussen says that there is one chance in 10 million that a reactor could lose its coolant, all backup systems fail and the radioactive material leak outside the plant sufficiently to cause 2,300 fatalities and $6 billion property damage. To illustrate the probability of such an accident, assume that at some future date we will have 500 nuclear power plants (60 nuclear plants were in operation in this country in 1976). Then, sometime in a 2,000 year period it is probable that there would be one accident of the severity described above. It may be argued that such a catastrophe should be avoided by banning nuclear power. However, it must be recognized that there is some element of danger in most of our activities. One person out of every 4,000 will be killed in an auto accident this coming year. Should we ban automobiles? To develop hydroelectric power, we construct large reservoirs. Many people have lost their lives when dams have failed. We send miners down in the mines to procure coal for us. Some of them will be killed in mine explosions. Others will contract the black lung disease. Should we stop using coal? It is possible that a very large airplane could crash into a crowded football stadium. The death toll could exceed that contemplated for practically any nuclear power plant failure. Should we abolish the airplane—or even football?

Another argument against nuclear power is that even when there are no accidents, the radiation from a nuclear plant has a marked effect on those people living near the plant. We are subjected to natural radiation from many sources. Rocks, such as granite, are quite radioactive. It was stated that the radioactivity from the granite walls of a Senate conference room was found to be as high as that in the immediate vicinity of the Three Mile Island plant right after the accident. We are subjected continuously to cosmic radiation and also radiation from terrestrial radioactivity. At high elevations, such as in Denver, Colorado, cosmic radiation is said to be as high as that adjacent to the Three Mile Island plant right after the accident. As individuals we were born radioactive. The food and water which we take in is radioactive. We pick up radioactivity from the roads over which we drive. It has been stated that the radioactivity from the products of a large coal burning power plant is many times that of a nuclear power plant. In general, if a person were to stand at the boundary of a nuclear power plant (about 7,000 feet from the reactor itself) all day for a year, he would receive less radiation than he would receive from one chest x-ray, by watching television four hours a day for a year, or by taking two long jet flights. Thus a careful examination of the facts should show that the radiation from a nuclear power plant has no significant effect on the local inhabitants.

Every feasible precaution must be taken to avoid nuclear accidents. The power plant must be carefully guarded. The radioactive material also must be carefully guarded when in transit. Means must be developed for storing the radioactive wastes safely. If all of these things are done carefully there is some chance, but a very, very small one, of serious consequences resulting from the use of nuclear power.

Recognizing the possible hazards connected with nuclear power production, we must make a very major decision. It has been projected that there will be a very large increase in the demand for electrical energy in the next two decades, not only to expand our present uses but also to supplant oil and natural gas. One source of energy for increased electric power production will be coal but a major portion of it is predicted to come from nuclear energy. Are we willing to reduce drastically the amount of electrical energy which we will use in the next two decades or are we willing to take some risk that appears to be a very small one, that nuclear power can be produced in desired quantities without serious adverse effects? As will be discussed in Part III of this text, it does not appear that any of the alternative energy sources will contribute much to our energy needs in the remainder of this century. Recognizing that a very slight danger may be involved, do we wish to develop the fast breeder reactor which promises to meet our energy needs for many centuries?

9.10 SUMMARY ON NUCLEAR POWER

The use of nuclear energy for electric power production is increasing quite rapidly in this country and also in the rest of the industrialized world. Approximately 9 percent of the electrical energy generated in this country in 1979 came from nuclear power plants. There were two reasons for this increase in nuclear power:

1. Nuclear power plants can deliver electrical power at a significantly lower cost than can fossil fuel plants, which means a substantial saving to the customers.
2. Although much electrical power is generated in oil or natural gas fired plants, these fuels are in short supply. Even though ample supplies of coal for power generation exist underground, there are very major problems both in mining and burning the coal without serious adverse ecological effects.

The known supply of uranium at economical prices is sufficient to last for two or three decades. The fast breeder reactor, now being developed, offers the potential not only of generating 50 to 100 times as much energy from a given amount of uranium but also the potential of using much more expensive and abundant uranium. It has been predicted that with the full development of the fast breeder reactor, nuclear energy can supply all of our energy needs for many centuries at a reasonable cost.

Unfortunately, there are potential hazards associated with nuclear power generation. These hazards involve the use of the nuclear fuel in the power plants, the transportation of both the fuel and the spent fuel elements to the reprocessing plant, and the disposal of the radioactive wastes. Elaborate safety precautions have been taken such that no one of the public to date has suffered seriously from nuclear power generation. To prevent serious accidents from occurring, more precautions and safeguards must be taken. In spite of all the precautions, it is possible, but very highly improbable, that some serious accidents may occur. Because of the possibility of these accidents, some earnest, knowledgeable individuals, together with many individuals having much less knowledge of the subject, are attempting to slow down or completely stop the use of nuclear energy. Thus, we must decide whether to proceed with development of this source which has the potential for abundant and relatively low cost power, even though there will be some minimal risks involved. The alternative seems to be a drastic decrease in the amount of electrical energy available in two or three decades from now.

SUPPLEMENT

PROBLEMS

9-1. Consider the use of nuclear energy for the power plant of Problem 6-1. Using the figure that the complete fission of one kilogram of U^{235} is equivalent to 2,400 tons of coal in producing electrical energy, determine the pounds of U^{235} required per week. (One pound equals 0.45359 kilograms.)

9-2. Determine the number of pounds of U_3O_8 required in Problem 9-1. Approximately 84.5 percent of U_3O_8 is uranium and 0.7 percent of uranium is U^{235}.

9-3. Uranium rock yields 5 pounds of U_3O_8 per ton of rock. Determine the number of tons of rock which must be mined per week for Problem 9-2. Compare this answer with the tons of coal required in Problem 6-1.

9-4. A major problem with nuclear power plants is that they require vast quantities of water for the condensers. Suggest how this problem may be minimized.

9-5. Some of the hazards connected with nuclear power production involve the transportation of the spent fuel elements. Suggest how this problem may be minimized.

9-6. Should additional safeguards be developed before any expansion can be made in nuclear fission power productions?

9-7. Should all expansion of nuclear production be halted?

9-8. Should this country proceed with the development of the fast breeder reactor?

9-9. Should any decisions be made as to the future of nuclear power production before an extensive study is made on the future of other present energy sources and the potential of alternative energy sources?

10

Hydroelectric Power

10.1 HYDROPOWER

Centuries ago, man learned to dam up streams and rivers to produce the head* necessary for power production. At first water wheels were used to develop the power. Most of the water wheels were either of the overshot or undershot type. In the overshot type, water flowed down on to the top of the wheel, somewhat off center, and produced work as it fell to a lower level. In the undershot type, water from the pond flowed down through a raceway, thus acquiring a high velocity. This high velocity water was then directed against the bottom of the wheel, causing it to rotate. The power produced by these water wheels was not converted to electrical power, but was used directly, particularly to drive factory machinery and to power gristmills for the grinding of grains. Although these hydroplants performed a useful purpose, by their very nature, the power output was rather limited.

The potential of hydropower became more fully utilized with the development of the hydraulic turbine in the early 1800's. The hydraulic turbine can handle very large quantities of water in a relatively (compared with the water-wheels) small space and, when desirable, water at very high heads. There are three types of hydraulic turbines: the impulse, the reaction, and the propellor type. For all cases, the turbine is located downstream at the base of the reservoir. In the impulse type of turbine, the water from the reservoir passes through nozzles and is directed onto the turbine rotor. In the nozzles, the potential energy of the water, due to its high pressure, is converted into kinetic energy which is utilized in turning the turbine rotor. The impulse type of turbines (known also as Pelton turbines) are used for water heads ranging from 300 to 5,000 feet.

In the reaction or Francis turbine, the water passes through converging passages within the rotor, thus having its potential energy converted into

* Head is the vertical distance between the surface of the water in the pond or reservoir and the bottom of the power producing device. See Figure 10-1.

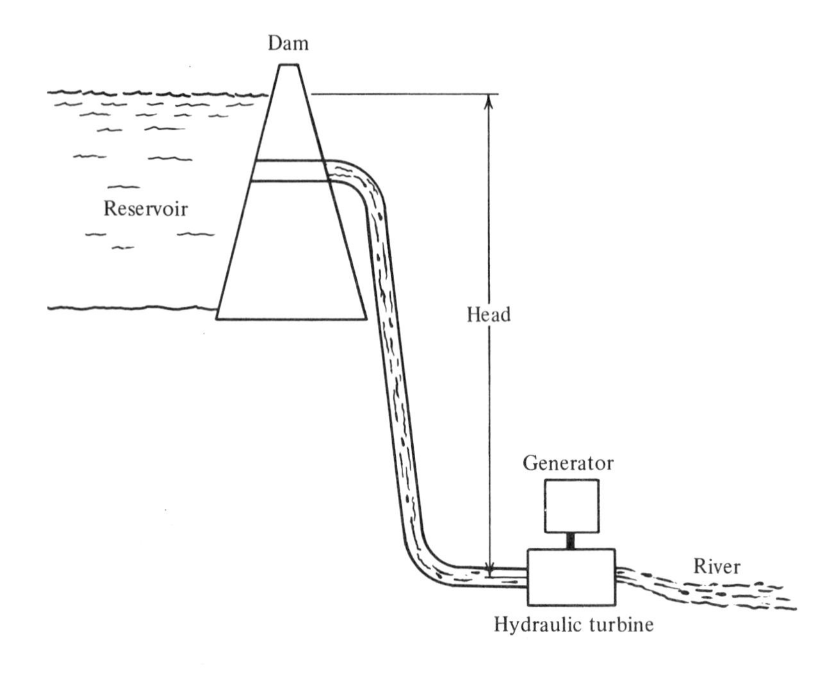

Figure 10-1. Element of a hydroelectric power plant

kinetic energy. During this conversion process, a reactive force is set up. Reaction turbines are used for heads varying from 40 to 1,000 feet. The rotor of the propellor or Kaplan type of turbine bears some resemblance to an airplane propellor. Water flows through nozzles, thus acquiring a high velocity. The water is directed onto the propellor blades where its kinetic energy is converted into energy of rotation. The propellor type of turbine is used for heads ranging from as low as 10 feet up to 150 feet. The efficiency of hydraulic turbines depends on their design and on how close they operate to their design conditions. These efficiencies range up to about 90 percent.

The theoretical power output of a water wheel or hydraulic turbine is given as

$$\text{Theoretical horsepower} = \frac{mh}{550} \qquad (10\text{-}1)$$

where m is the flow of water per second, pounds
h is the head in feet
550 is the ft. lb. per hp. sec.

The efficiency (η) of power production is the ratio of the actual power to the theoretical power, or

$$\eta = \frac{\text{actual power}}{\text{theoretical power}} \qquad (10\text{-}2)$$

Then:

$$\text{Actual horsepower} = \frac{mh\eta}{550} \tag{10-3}$$

When the volume rate of water flow is given in cubic feet per second, the value of m can be obtained by multiplying the volume rate by 62.4 lb. per cubic foot Sometimes the volume flow rate is given in gallons per second. The value of m can be obtained by multiplying the gal. per sec. by 8.34 lb. per gal.

Examination of Equation 10-3 shows that vast quantities of water are required per second even though the head is quite high. Thus, large hydroelectric power plants require rivers having high rates of water flow and a region permitting the construction of reservoirs to establish the desired head.

Years ago many small hydro plants were constructed throughout the country. Streams were impounded to create reservoirs. Generally, water wheels were employed to produce power for such purposes as grinding grain or operating small factories, particularly textile mills. Many of these hydro plants were abandoned when factory operations were concentrated in much larger operations at other locations. Cheap electricity made this not only feasible but desirable since electric motors provide flexibility not possible with direct drive from water wheels.

With the large increases in the cost of electric power, consideration is now being given to reclaiming these abandoned hydro plants to produce electric power. This seems to be economically feasible in many cases, particularly where the dam is still in existence and in good condition together with its reservoir. The Corps of Engineers has estimated that the national hydroelectric power potential at existing dams for capacities of less than 5,000 kW and heads of less than 100 feet is 26.6 million kW. It is difficult to predict how much of this potential power can be economically developed. The Department of Energy (DOE) is aiding in the funding for some of this development. However, several problems exist. The present dam may require extensive repairs. The installation of a hydraulic turbine and its electric generator is much more complex than that of a water wheel. The initial cost per rated kilowatt is high for smaller units. The costs for plant operators is almost as large for small plants as for larger ones. In some cases when dams were abandoned years ago, the reservoir dried up and the land was used for other purposes. Nevertheless, because of the need for additional electric power sources, all abandoned hydro power systems should be examined carefully.

10.2 PRESENT AND FUTURE CAPACITIES

The first hydroelectric central station was started in Appleton, Wisconsin on September 30, 1882. The growth of hydroelectric power was relatively slow at first. However, in the early 1930's, hydroelectric power plants supplied almost one-third of our electrical energy. Since that time there has been a large development in this field throughout the country, particularly in the Tennessee

River Valley, the Colorado River Valley, the Columbia River Valley and the Missouri River Valley. But in spite of this great growth of hydroelectric power, our demands for electricity have grown so rapidly that hydroelectric power now supplies only about 13 percent of the energy for our electrical needs.

Table 10-1 shows the present hydroelectric power generating capacity and also the potential generating capacity for various regions of the country. It is rather difficult to estimate the amount of hydroelectric power potential since the magnitude of the head which can be created is a function of the amount of land to be flooded to create the reservoir. Although it is possible to produce hydroelectric power with somewhat lower heads, it is generally assumed that at present the lowest economic head is 20 feet for large scale power production.

Table 10-1. U.S. Hydroelectric Power Resources

Region	Developed Capacity Megawatts	Potential Power Megawatts	Percent Developed
New England	1,500	4,800	31.3
Middle Atlantic	4,200	8,700	48.3
East North Central	900	2,500	36.0
West North Central	2,700	7,000	38.0
South Atlantic	5,300	14,800	35.8
East South Central	5,200	9,000	57.8
West South Central	1,900	5,200	36.5
Rocky Mountain	6,200	32,900	18.8
Pacific	23,900	62,200	38.4
Subtotal	51,800	147,200	35.2
Alaska	100	32,600	0.3
Hawaii	0	0.1	0
Total	51,900	179,900	28.4

This table was taken from *Energy Alternatives, A Comparative Analysis*, prepared for various government agencies by The Science and Public Policy Program, University of Oklahoma, Norman, Oklahoma, May 1975.

10.3 PROS AND CONS OF HYDROELECTRIC POWER

Hydroelectric power is a renewable source. As such, it may be expected to supply energy indefinitely. However, particularly in certain regions, much silt is carried into reservoirs, limiting their useful life. In some regions, land to be flooded to create a reservoir is wilderness and is relatively cheap, but in other regions, the land may be very expensive. The amount of land required for a given head and a given power plant output varies greatly with the topography of the region as does the size of the dam. Hence, it is very difficult to establish a figure for the cost of a hydroelectric power plant per kW output. In general, though, the cost of hydroelectric power is much lower than power generated in fossil fuel plants. The cost of power production from government owned hydroelectric plants is generally very low since the interest rate on the money required to build the plant is low as are taxes or their equivalent.

In addition to providing electric energy, reservoirs created often serve the useful purpose of flood control. It is difficult to place a monetary value on the savings in property and life made by creating reservoirs, but some of the original cost of the dam and reservoir may be logically charged to flood control, thus reducing the cost of power production. Depending on the regions in which they are located, reservoirs also can be used for recreational purposes. Again it is difficult to assign a monetary value to this use.

Although, as can be seen from Table 10-1, there is much potential for future hydroelectric development in all regions of the country, it is difficult to predict the extent of this development. Much of the land required for the reservoirs is river bottom land which is rich and productive. It is the home, in many cases the ancestral home, of people who are very adverse to leaving. In some parts of the country, the rivers which may be used for hydroelectric power generation flow through scenic regions. There is very strong reaction from environmental groups to damming these rivers and thus destroying the scenic wilderness.

There is some fear of dam failure. There have been failures over the years, particularly smaller ones, but also larger ones. In June of 1976, a 320 foot high, newly constructed dam on the Teton River in Idaho failed, causing loss of life and perhaps over a half billion dollars in monetary damage. It is difficult to predict the probability of dam failure and the extent of damage which may result. The probability, however small, is there, just as it is for nuclear power plants.

Decisions concerning further development of hydroelectric power must involve consideration of several important factors. Are we willing to destroy scenic areas, flood productive farm lands and displace their inhabitants? Do we want to take a chance, however small, that a dam failure could cause a serious loss of life? Or are we willing to so decrease our demand for electrical energy that further expansion of hydroelectric plants will not be necessary?

10.4 PUMPED-STORAGE

Pumped-storage does not create energy but is simply a means of storing energy which has already been generated. This is accomplished by pumping water during periods of low power demands to an elevated reservoir. During periods of high power demands, the water flows back through a turbine to produce hydroelectric energy to help meet the high electric power demands. (See Figure 10-2.) Thus far in the text consideration has been given to production of energy from natural sources. Pumped-storage is included here only because many of the problems involved are similar to those for hydroelectric power.

For most electric power systems, there is a wide variation in the power demands not only throughout the day but also week, month and year. The

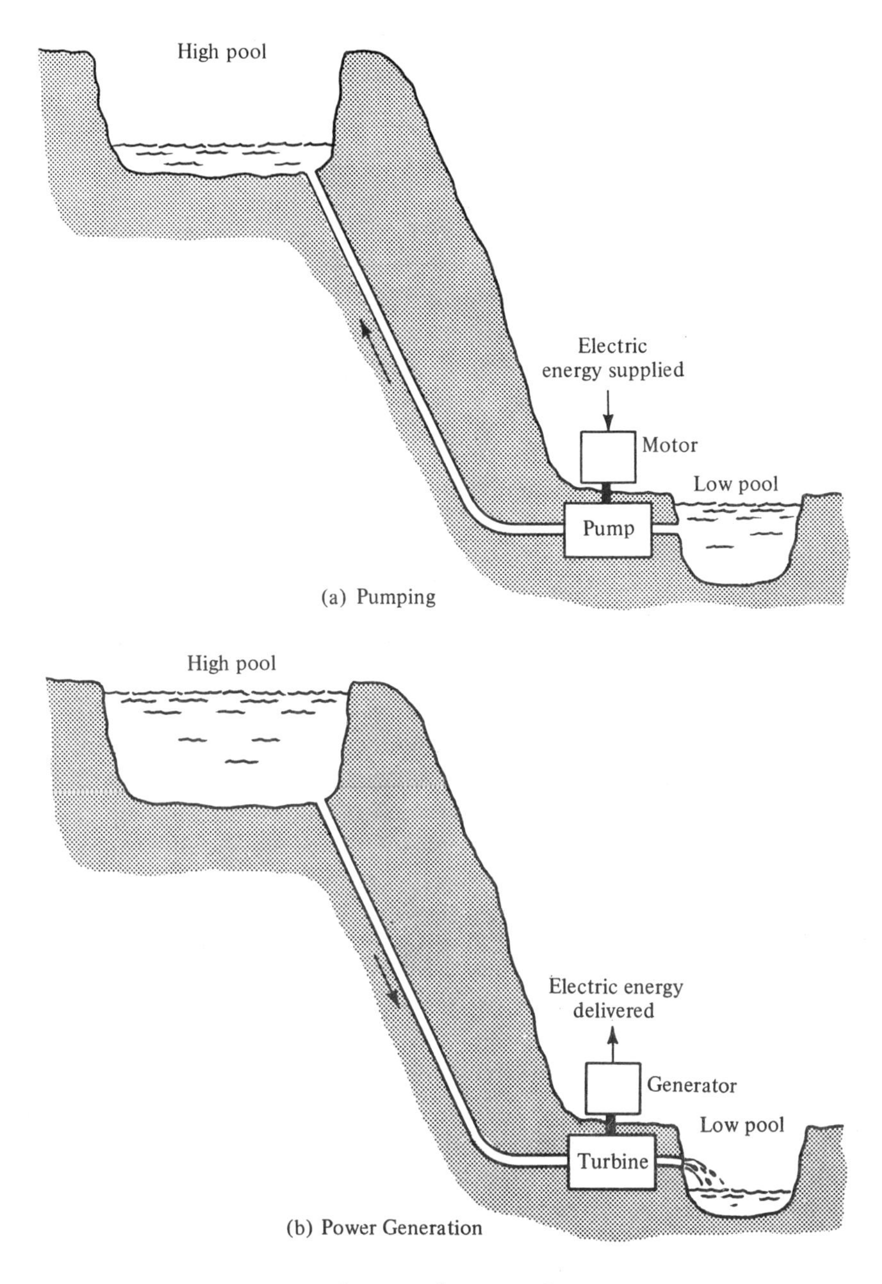

(a) Pumping

(b) Power Generation

Figure 10-2. Elements of a pumped storage system

power plants for a given system must be sufficiently large to meet the maximum demand at all times. This can be accomplished by installing enough units to generate the maximum possible amount of power which may be demanded. However, this would mean that large expensive equipment would be idle a large portion of the time. As discussed in Chapter 4, the fixed charges on this idle equipment greatly increases the cost of power production.

Pumped-storage requires a very large source of water at a comparatively low elevation, which may be used for pumping purposes. The withdrawal of large quantities of water from the source must not cause adverse environmental effects on the source, such as interfering with aquatic life or making an excessive decrease in downstream flow. In addition, space is required at a high elevation, close to the water source, which is suitable for making a reservoir. If a significant amount of power is to be produced, the reservoir must be very large. These three requirements, a large supply of water suitable for pumping, an elevated area to produce a large head, and sufficient area at a high elevation to permit the construction of a large reservoir, greatly limit the extent to which pumped-storage is feasible.

A specially designed unit may be used in connection with pumped-storage which can be used as a pump during the pumping period and as a hydraulic turbine during the power producing period. The overall efficiency of a pumped-storage system is somewhere around 70 percent although the efficiency in some installations is much lower. Whereas this overall efficiency may appear to be good, it must be understood that almost one-third of the electrical energy generated in the conventional power plant for this purpose is lost. This means that more fuel, fossil or nuclear, must be burned to deliver a given amount of power to the customers.

To explain why it is economically desirable to use pumped-storage, assume that a power company anticipates that 500,000 kW of extra power producing capacity will be required to meet peak load requirements. Assume that the three conditions, discussed earlier, are fulfilled to permit a pumped-storage installation. To meet this demand the power company can install a conventional 500,000 kW power plant. The cost of this plant will exceed the costs of a 500,000 kW pumped-storage plant. Although the conventional plant will require less fuel than will be used overall for the pumped-storage plant, the overhead costs of the conventional plant generally will more than offset the added fuel costs for the pumped-storage plant. An alternate solution would be to install 500,000 kW of gas turbines. The gas turbines are much cheaper than the conventional units but are much less efficient and generally use expensive fuel. It is possible that the cost of fuel when gas turbines are used to carry the peak load will exceed the total fuel costs for pumped-storage. Although many factors are involved when pumped-storage is being considered for handling peak loads, in many cases it is economically desirable to install pumped-storage.

The environmental objections to pumped-storage plants are similar to those for conventional hydroelectric power plants. The drawing of vast quantities of water from a river or lake could affect the aquatic life and, in the case of a river, materially affect the downstream river flow. Much land would be required for a reservoir. However, unlike the hydroelectric plant, the pumped-storage plant would not be very well suited for recreational purposes since it would be full at the start of the day and well drawn down at the end of the day.

The possibility of expanding pumped-storage plants must be considered. Here, the savings in the cost of peak power production should be balanced against possible adverse environmental effects. Because this decision will affect the general public, who will be paying the bill, the general public should have a major input to the final decision.

10.5 SUMMARY ON HYDROELECTRIC POWER

Hydroelectric power plays a significant role in meeting our demands for electrical energy, supplying about 13 percent of these needs. As shown in Table 10-1 in certain regions of the country much reliance is placed on hydroelectric power. Because hydroelectric power is a renewable resource, it should be used wherever feasible. Table 10-1 shows that it is possible to make a large increase in the hydroelectric installations in this country. However, it is anticipated that our demands for electrical energy will grow much faster than new hydroelectric plants can be installed and hence hydroelectric plants will supply an ever decreasing portion of our electrical energy. Not only do hydroelectric plants produce electrical energy, but in many cases, their reservoirs aid in flood control and also provide recreational facilities. On the other hand, the reservoirs require large tracts of land, often scenic land or valuable farm or industrial land. There is some possibility of dam failure which could cause serious loss of property and life.

In those areas of the country where large quantities of water are available for pumping, and where it is possible to create a large reservoir at a nearby high elevation, it is anticipated that there will be an increase in pumped-storage installations. In spite of the losses in electrical energy in such a system, the lower fixed charges as compared with a conventional plant make it a desirable way to take care of peak loads on a power system. When feasible to use a pumped-storage system, power costs to the customers can be held down.

SUPPLEMENT A

ILLUSTRATIVE EXAMPLES

Example 10-1. The mean flow of a river is 800 cu. ft. per sec. (cfs). A dam can create a head of 120 ft. Assume an efficiency of 88 percent; determine the average power output.

Solution. The mass rate of flow is $800 \times 62.4 = 49{,}920$ lb. per sec. (62.4 is the density of water, lb. per cu. ft.).

From Equation 10-3,

$$\text{hp} = \frac{mh\eta}{550} = \frac{49{,}920 \times 120 \times 0.88}{550} = 9{,}585 \text{ hp} \qquad \text{answer}$$

Example 10-2. The river in Example 10-1 has an average depth of 3.5 ft. and an average water felocity of 2 ft. per sec. Determine its width.

Solution. The product of width, depth, and velocity equals the volume flow per unit time. Then

$$\text{width} = \frac{\text{volume flow}}{\text{depth} \times \text{velocity}} = \frac{800}{3.5 \times 2} = 114.7 \text{ ft.} \qquad \text{answer}$$

Example 10-3. A pumped-storage power plant is to deliver an average of 250,000 kW for an eight hour period. The mean head is 84 feet. The overall efficiency is 68 percent. Determine the volume of water which must be stored.

Solution. From Equation 10-3, the amount of water required, on the average, per second is:

$$m = \frac{250{,}000}{0.746(84 \times 0.68)} 550 = 3{,}227{,}000 \text{ lb. per sec.}$$

where 0.746 = kW per hp

$$\text{Volume} = \frac{3{,}227{,}000}{62.4} = 51{,}710 \text{ cfs}$$

$$51{,}710 \times 3{,}600 \times 8 = 1{,}489{,}000{,}000 \text{ cu. ft.} \qquad \text{answer}$$

SUPPLEMENT B

PROBLEMS

10-1. Determine the volume rate of flow required to produce 100,000 kW. The head is 185 ft. The efficiency of the turbine is 90 percent.

10-2. The dam in Example 10-1 creates a reservoir covering 12,200 acres. The average depth of the reservoir is 65 feet. Determine the volume and the mass of the water in the reservoir.

10-3. The reservoir for Example 10-3 has an average depth of 25 feet. Determine the surface area required in acres.

10-4. Repeat Problem 10-1 for heads of 25 ft. and 370 ft. Make comments on the effect of head on the amount of water required.

10-5. Comment on what could happen if the reservoir in Problem 10-2 should break.

10-6. If a pump was used for pumping purposes and a separate turbine was used for power producing purposes in a pumped-storage system, there would be some gain in efficiency. Would you recommend this? Explain.

10-7. Do you feel that there should be a national policy regarding the installation of additional pumped-storage plants? If so, what should it be?

10-8. Should it be anticipated that the full potential of hydropower as given in Table 10-1 will be almost all developed ultimately? Why?

10-9. Determine the average power potential in kW of an abandoned hydropower plant. The average velocity of the water in the river feeding the reservoir is 2 feet per second. The average river width is 24 feet and the average depth is 2.8 feet. The average height for power production is 5.2 feet and the average overall efficiency is 88 percent.

10-10. The river flow in Problem 10-9 averages 50 percent of the normal for a period of four months. Water is to be stored in a reservoir to provide a total normal flow to the turbine during this period. The average depth in the reservoir is 4.5 ft. during this time. Determine the area of the reservoir. Is this practiced?

PART III

OUR ALTERNATIVE ENERGY RESOURCES

11

Introduction

Substantially all of our energy is presently obtained from fossil fuels, nuclear fuel and hydroelectric power. There are many other energy sources which offer potential for supplying a significant portion of our energy before the year 2000. Even though some of these energy sources have been used to a very limited extent in the past, they do not presently make a significant contribution to our energy supply. Hence we will consider them to be alternative energy resources.

The discussion of these alternative energy resources will consider the potential of each. Some of these sources, such as the sun, the wind, and the oceans, are renewable resources. Other sources, such as tar sands and shale oil, are non-renewable resources and hence are of limited potential when viewed over the long term. It should be noted that, with the exception of geothermal energy and fusion, all of these energy resources come directly or indirectly from the sun.

Technology now exists for using most of these new energy resources but in general it has not been economically attractive to develop many of them. For some, though, improvements in manufacturing techniques and mass production may so reduce costs that it will become economically sound to utilize these resources, particularly as the costs of our present energy supplies increase.

The technological feasibility of many of these alternative energy resources has been demonstrated analytically, in the laboratory and/or in very small installations. The economics of full-scale installations have been projected from these small scale operations. Detailed estimates have been made for full-scale size units in many cases. However, all such extrapolations should be questioned. Frequently these estimate show that the full-scale unit will be economically sound. However, final costs may prove to be two or three times the estimated cost, completely negating the anticipated advantage of using a particular energy source. It is true that the necessary development costs make the first few units of any new system quite expensive. It is very difficult to predict how much reduction in costs can be made after development costs are written off.

It is even more difficult to predict that major technological breakthroughs can be achieved which will make possible the economic feasibility of using a particular energy source. With so much uncertainty in the possible costs of energy to be provided by these new sources, the question must be asked, should there be a national policy concerning the use of these new energy sources and, if so, what should it be?

On the one hand, there are those who advocate spending large sums of money, using tremendous amounts of our resources to develop full-scale systems to use alternative energy sources, in the hopes that these systems may prove to be effective. On the other hand, there are those who feel that very little effort should be expended on these energy sources but extensive efforts should be made to expand our use of coal or to develop the fast breeder reactor, since such efforts will provide us with the desired amount of energy for many centuries.

Perhaps a more logical approach would be one between these extremes. Because of the high potential that many of these sources have for supplying us with a significant portion of our energy, it seems reasonable to proceed step-by-step in investigating the possibilities of developing these energy sources. It is very risky to proceed directly from laboratory models to full-scale plants. Rather, pilot plants should first be built to provide both operating and cost information. If this information is favorable, somewhat larger and then, finally, full-scale plants can be built. Only from information obtained on full-scale plants will it be possible to estimate initial costs of full-scale plants to be mass produced. In addition, extensive research and development work should be undertaken to try to make breakthroughs in the technology and in the costs of using these alternative energy sources.

Many estimates are available of the projected costs of using these energy sources. Some of these estimates show such a definite economic advantage for a particular source that it appears that all efforts should be made to develop this source as soon as possible. On the other hand, other estimates of the cost of producing enegy from the same source indicate that it is doubtful that it should ever be used, at least in the next decade or two. Because of this great uncertainty in predicting costs of energy from these sources, such varying estimates will not be presented here, since they could be confusing and misleading to the reader. Rather, the technology of utilizing each alternative energy source will be discussed here, including both the problems involved as well as the potentials. Furthermore, consideration will be given to the economics of each process as it is developed step-by-step in order that its feasibility can be evaluated.

12

Direct Solar Energy

12.1 THE SOLAR ENERGY RESOURCE

Even though the earth receives only a very small fraction of the energy radiated by the sun, the solar energy which arrives at the earth's surface is tremendous. It has been estimated that the earth receives 2.4×10^{21} Btu of solar energy per year, which is approximately 18,000 times the amount of energy used by all man-made devices throughout the world in 1970. What happens to this energy? Much of it is either reflected or radiated back into space, this amount of energy being dependent on the nature of the surface receiving the radiation.

Since approximately 70 percent of the earth's surface is covered by water, land areas receive approximately only 30 percent of the solar energy striking the earth. Some of the solar energy falling on the oceans is used to vaporize water which gives us our rain. Much of the incident* solar energy warms the oceans. One result of the solar heating of the oceans is the generation of various currents, such as the Gulf Stream. In turn, the Gulf Stream warms those European countries bordering on the Atlantic Ocean to temperatures significantly above those normally associated with regions that far north. Because the surface temperature in many parts of the oceans is much higher (27 to 35°F) than that a half or so mile down, the potential exists for power generation. This will be discussed in Chapter 15.

On the land, some of the incident solar radiation is absorbed by plant life and, indirectly, by animal life. This energy thus absorbed for millions of years created our fossil fuels. This process is still going on in the formation of peat in such places as Ireland and the northern plains states in this country.

Wherever it strikes, solar energy has a tendency to produce a warming effect. As the temperature of the objects receiving solar energy increases, there is an increase in the energy given off by these objects. By this process much of the

* Incident solar energy is the solar energy striking the earth.

total energy received by the earth on both the land and water areas heats the air above the surface, thus producing air currents and the winds.

The amount of solar energy received by a given area is dependent on the location of that area, the time of year and the extent of cloud coverage. Table 12-1 gives the solar radiation received at cities in various locations throughout the United States.

Table 12-1. Average Total Daily Insolation (Btu per square foot per day)

Location Month	Fresno, California	Tucson, Arizona	Omaha, Nebraska	Lakeland, Florida	Atlanta, Georgia	Burlington, Vermont
January	710	1,110	777	1,029	873	581
February	1,117	1,391	1,110	1,436	1,203	781
March	1,709	1,750	1,284	1,480	1,288	1,088
April	2,205	2,202	1,576	1,983	1,635	1,384
May	2,609	2,435	1,939	2,079	1,991	1,447
June	2,579	2,442	2,165	2,042	1,854	1,758
July	2,576	2,190	2,002	1,883	1,917	1,587
August	2,412	1,983	1,865	1,680	1,628	1,835
September	2,050	1,735	1,280	1,639	1,591	1,195
October	1,425	1,587	944	1,436	1,021	755
November	910	1,221	581	1,302	955	444
December	610	870	596	1,169	714	448
Average	1,743	1,745	1,351	1,597	1,389	1,109

This table was abstracted from *Energy Alternatives—A Comparative Analysis*, prepared for various government agencies by the Science and Public Policy Program, University of Oklahoma, Norman, Oklahoma, May 1975.

It has been estimated that the average insolation* for continental United States is 1,450 Btu per square foot per day. As shown in Table 2-1, it is predicted that our daily energy demands in 1990 will be 0.278 Quads per day. Thus if we could fully utilize the total solar energy falling on an average area, then the area required to collect the energy to meet our contemplated needs in 1990 would be 4.40×10^6 acres or 6,878 square miles. This figure is deceptively small. First of all, a significant portion of the incident solar energy will be either reflected or re-radiated. In addition, the efficiency of utilization of solar energy for power production is low. As will be discussed in this chapter, we possess the technical knowledge to use solar energy for electric energy generation. However, present indications are that the efficiency of this electric energy generation may be less than one-half of that of our best fossil fuel plants. This means that at least twice as much solar energy must be collected for a given electrical output. At present, most of our energy needs, other than electrical needs are satisfied by oil or natural gas. It is possible to use solar energy, indirectly, to produce a synthetic fuel, such as hydrogen, to replace oil and

* Insolation is the rate at which the earth receives solar energy, normally expressed in Btu per square foot per day.

natural gas. For instance, solar energy could be used for electric power production and hydrogen then produced by electrolysis of water. However, the overall efficiency of such a process would be low, perhaps not more than 10 percent. This means that 10 times as much solar energy would have to be collected than would be present in the hydrogen produced.

Summing up, it seems that a collecting area of at least 100,000 square miles would be required if we were to rely entirely on solar energy to satisfy our needs in 1990. This means that it would be necessary to use a collector area equal to the total area of Delaware, Maryland, North Carolina and Virginia. These figures are presented here to illustrate the point that solar energy is very diffused and therefore vast areas are required to collect significant portions of it. Thus, although solar energy is free, the collecting of it is inherently expensive.

Although solar energy may be used in various ways, the three ways now receiving the most attention are

1. Direct heating and cooling.
2. Direct production of electrical energy.
3. Indirect production of electrical energy.

12.2 DIRECT HEATING AND COOLING

Solar energy has been used for various heating purposes, such as cooking, crop drying, fresh water production and salt production from ocean water. However, most of the efforts to use solar energy for heating are being directed towards

1. Metallurgical purposes.
2. Hot water heating.
3. Space (building) heating and cooling.

1. Solar Furnaces. By using focusing collectors, it is possible to attain furnace temperatures as high as 5,500 to 6,000°F. Such a temperature is most desirable for certain metallurgical purposes. Furthermore, this heating can be carried out without fear of contamination from a fuel or electrical heaters or electromagnetic fields. It is relatively easy to control the reactions as well as the temperatures.

A solar furnace was used as early as 1774 by Lavoisier for carrying out chemical studies at high temperatures. In 1957, 21 solar furnaces were in operation in the United States. Large solar furnaces were also built in the U.S.S.R., Japan, Algeria, France and other countries. Some years ago, a very large solar furnace was constructed in Odeillo, France. Parabolic mirrors covered a seven-story building to collect and focus the solar energy into the metallurgical furnace. Temperatures approaching 6,000°F were attained. The solar energy input was equivalent to approximately 1,000 kW.

The high initial costs of solar furnaces have been a barrier to their wide use, particularly since until rather recently, natural gas and electrical energy costs were relatively low. Now with the increasing costs of these energy sources, more attention is being devoted to solar furnaces, in particular on developing less expensive parabolic concentrating mirrors. Both plastics and specially treated aluminum are being considered for this purpose. Solar furnaces are valuable for specific uses, but the total energy used by them is only a small percentage of our national use. It does not appear that solar furnaces will ever make a significant reduction in our energy demands on other sources.

2. Hot Water Heating. Approximately four percent of our total energy supply is used for the production of hot water, residential and commercial. Although this is not a large portion of our total energy use, if solar energy could be used to supply this heating, it could make a significant effect when combined with substitutions of solar energy in other areas and with real energy conservation efforts. A solar hot water heating system consists of the collector, the energy transfer system and a storage tank to provide hot water when the collector does not receive sufficient energy to meet demand.

The heart of the solar water heating system is, of course, the collector. Various types of solar collectors have been designed and used. One of the simplest types is the flat-plate collector. A version of this collector is shown schematically in Figure 12-1. As the name implies, it consists of a flat-plate which is blackened to absorb the solar radiation. To minimize heat losses, it is covered with two panes of glass. Tubes are attached intimately to the collector surface. A fluid, such as water or air, is circulated through the tubes to pick up and transport the heat which has been absorbed by the collecting surface. Heavy insulation minimizes heat losses from the back of the collecting surface.

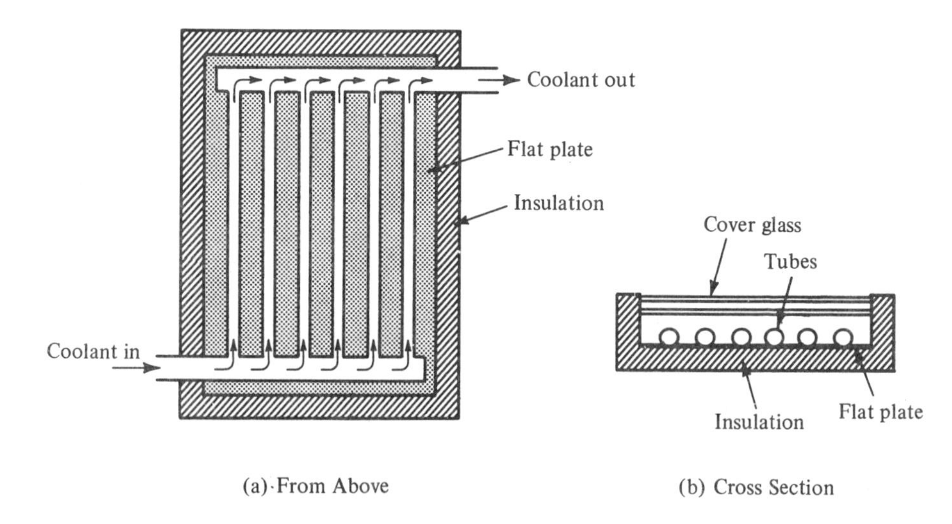

Figure 12-1. Flat plate collector

Several factors must be considered in the design of the collector. The major factors are

a. Collecting the solar energy.
b. Minimizing the loss of the collected energy.
c. Transferring the collected energy.
d. Deterioration of the heat transfer elements and the resulting possible contamination.
e. Possible freezing when water is used as the heat transfer agent.

a. Collecting the Solar Energy. A dull, black surface absorbs almost all of the solar radiation which it receives. This fact dictates the nature of the absorbing surface. In addition, in order for the absorbing surface to receive the maximum amount of solar energy, it should be mounted so that it is perpendicular at all times to the rays of the sun. This means that there should be some mechanism to automatically adjust the angle of the collector. However, the cost of such a mechanism is presently too high to justify its use. Thus an overall optimum angle at which to position the collector surface must be determined. In the winter, water must be heated over a large temperature range while the intensity of the solar radiation is at its lowest. (See Table 12-1.) Hence, more consideration should be given to winter operations in selecting the angle of the collector. As a compromise, the collector should face south, with the angle between the collector and the horizontal being greater, perhaps as much as 10 degrees, than the angle of latitude for the given location.

It is common practice to locate the collector on the roof of the building. Consideration must be given, however, to possible shading of the collector either by adjacent buildings or by trees. It should not be forgotten that trees grow significantly and may shade the collector before the end of its useful life. In order to keep the collector unshaded, it may be necessary to mount it on a platform placed on the ground. However, this location may cause an increase in the cost of the piping system to transfer the heated fluid from the collector.

b. Minimizing Energy Loss from the Collector. By virtue of its temperature, the collector surface tends to lose heat, from radiation and from the heated air which is in direct contact with it. This heated air tends to transmit heat to the surroundings. This is particularly true if the wind is allowed to blow over the collector surface. Losses can be minimized by placing one or two panes of glass over the collector surface. The panes of glass are separated from the collector surface and from each other by a dead air space to minimize the heat loss from the heated air which is in direct contact with the collector surface.

The wavelength of radiant energy varies with the temperature of the surface from which it is emitted. The wavelength of solar energy is short. As such, it readily penetrates the glass plates and impinges on the collector. Since the temperature of the collector is relatively low, the wavelength of the energy which it radiates is long, and hence it does not pass readily through the covering glass plates. Plastics have been considered in place of glass for cover plates since some plastics cost substantially less than glass. However, most plastics do not transmit solar energy as readily as glass. In addition, they have a tendency to deteriorate when exposed to heat and storms. It is not possible at this time to predict whether or not satisfactory plastics can be found to replace glass.

It is not difficult to prevent significant heat losses from the rest of the collector. Relatively cheap insulation, such as fiberglas, is available for this purpose. However, unless care is taken in design, there may be some heat losses through the structural supports.

c. Transferring the Collected Energy. Air has certain advantages as a fluid for transferring energy from the collector. When it is used, the problem of freezing is eliminated. Although it may oxidize the heat transfer surface, it will not corrode as water can. However, air is a poor heat transfer agent. Larger pipes or ducts are required to transfer it. In addition, a heat exchanger is required to produce domestic water. For these reasons, air generally is not used for solar heating hot water systems, although it may be used for solar building-heating systems.

In some ways, water is a good heat transfer agent. The pipes required for its transfer are small. If it is not contaminated in the collector, it may flow directly into the domestic hot water tank, thus eliminating the need for a heat exchanger. To prevent contamination in the collector, it may be necessary to use copper, which is expensive and in rather short supply.

d. Deterioration of the Heat Transfer Surfaces. There are plastic materials which may be used for the tubes of the collector which do not deteriorate significantly and which will not contaminate the water. However, plastics are such poor heat conductors that it is doubtful if they can be used for this purpose. Materials which can be considered are copper, aluminum, galvanized steel, and stainless steel. Stainless steel is very expensive and hence its use is difficult to justify. Assuming that water is the heat transfer fluid, after some years of use, galvanized steel would tend to deteriorate and thus contaminate the water. Although some aluminum alloys resist corrosion, many authorities feel that there would be serious deterioration of the aluminum tubes in eight to ten years. Before this time, there would be some contamination of the water. In addition, aluminum has a tendency to interact badly with copper, which is commonly used in residential hot water heating systems. Copper is, in most respects, an excellent tube material. However, it is expensive and will be in short supply. It is felt by some that there will be a severe shortage of copper to meet our ordinary needs by the year 2000. If large amounts of copper were to be used for solar collectors, the shortage would become more severe and the cost greater. To prevent contamination of the domestic hot water, a heat exchanger can be installed either in the storage tank or between the storage tank and the collector. This would permit using a material cheaper than copper for the tubes in the collector, but the cost of the heat exchanger together with the increased complexity of the system may offset the cost gain by substituting for the copper.

e. Freezing of the Water. In a large portion of our country, freezing could occur in the winter if water were the heat transfer fluid. In those regions of the country where freezing occurs only occasionally, the system may be designed to be drained for the relatively small number of times when freezing is expected. In colder regions, the only practical solution seems to be the addition of anti-freeze to the water circulating through the collector. If this is done, the water could not be used for domestic purposes and a heat exchanger would be required.

Since, even in clear weather, the sun shines only part of the time, a hot water storage system is essential. A large heat storage system as well as a large collector will be required when solar energy supplies all of the heating. Although the collector absorbs much of the diffuse or sky radiation, even when the sun is not shining, unless the collector is tremendous in size, it cannot pick up sufficient energy in cloudy, cold weather. Furthermore, it is not economically feasible to store enough hot water for a protracted cold, cloudy period. Hence an auxiliary system must be provided for supplying hot water during these periods. To determine the size of the water storage tank required, it is necessary to know how much hot water will be required per day and for what period it is to be stored. A figure of 15 to 20 gallons of hot water per person per day is commonly used to estimate the necessary storage capacity. Recognizing that some heat will be absorbed by the collector, even on a cloudy day, perhaps an 80 gallon storage tank may be satisfactory for a family of four. Such a system should be ample for perhaps 60 to 80 percent of the time. The size and cost would have to be doubled or tripled to eliminate the need of an auxiliary heating system.

Space being available, it is possible to mount the storage tank above the collector. As the water becomes heated in the collector, it will rise by buoyant action to the top of the tank, to be replaced by cold water from the bottom. If such an arrangement is not feasible and it becomes necesary to locate the tank at a level below that of the collector, a pump will be required. Such a system is shown schematically in Figure 12-2.

The size of the collector will also depend on the percentage of the total time it is to supply hot water as well as on its efficiency. Some figures that have been given indicate that the efficiency of the best collector may be as high as 70 percent. The efficiency of the collector decreases as the temperature of the water it produces increases. If it is to deliver water at 140° to 150°F, the efficiency of the average collector may be as low as 45 to 60 percent. The efficiency and hence the amount of heat collected by a given collector can be increased somewhat by decreasing the temperature of the water it produces to perhaps 125° to 130°F. However, a larger storage tank would be required, not

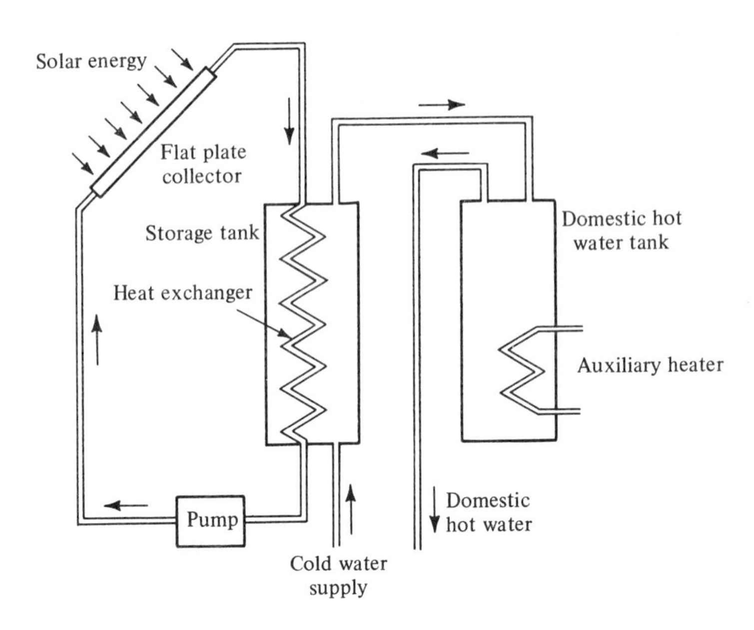

Figure 12-2. Solar hot water heating system

only because there would be somewhat more energy to store but also because each pound of water contains less energy when its temperature is reduced. A figure of 60 to 80 square feet has been suggested as a minimum size of the collector for a family of four.

It is rather difficult to estimate the initial cost of a solar hot water heating system, since there are so many variables involved. It is possible, in existing homes, that the hot water tank now being used may be of sufficient size for the solar system. If the tubes in the collector are copper, no intermediate heat exchanger may be required, particularly in regions where severe freezing may not occur. Under these conditions, only the collector, a pump, and the piping system may be required. Even when the householder does not make his own installation, there is a wide range in the estimated costs of a simple solar hot water heating system. Although costs as low as $400 have been given, it seems that a realistic figure may be as high as $1,000. The cost for a more elaborate system, containing automatic controls, will be much higher.

In general, it appears that the use of solar hot water heaters is economically sound in those warmer regions where energy is expensive. In warmer regions, the average solar insolation is relatively high, as is the temperature of the water to be heated. At present, the regulated price of natural gas is so low that it may be preferred for heating water. Solar hot water heaters are now used to some extent in Florida and California in this country and in foreign countries, particularly Japan, Australia, Israel, and the U.S.S.R. Based on increasing energy costs, it is anticipated that there will be a very large expansion in the use of solar hot water heaters.

3. Building or Space Heating and Air Conditioning. Many of the factors involved in the use of solar energy for space heating are similar to those involved in the solar production of hot water. It is suggested that the reader review the section dealing with solar hot water heating before proceeding with the discussion of space heating. In spite of the similarity, there are some large differences in the two systems, namely in the size of the collector and also the storage system. In addition, it is possible to adapt the solar space heating system to provide air conditioning in the summer.

There is wide variation in the amount of heat which will be lost from any building. The heat lost will depend on the size and type of building, how well it is insulated, where it is located, the outside weather conditions, the heat given off by the activities within the building, and also the temperature to be maintained within the building. These variables, coupled with the percentage of time which solar energy is to carry the heating load, makes it most difficult to give even an approximation as to the size of the collector required. In general, the collector area seems to range from 40 to 75 percent of the floor area of the building to be heated.

Either water or air may be used to transfer heat from the collector. Freon is also being considered for this purpose, if a heat pump* will be used in collection with the system for summer air conditioning. The choice of the heat transfer agent is related to the type of the heat storage system to be used. In many ways, rocks make a good heat storage system. They are inert and, in many regions, plentiful and cheap. With rocks as the storage system, air is a convenient heat transfer agent. Although it is not a good heat transfer agent and does require larger pipes than does water, it will not cause the corrosion which water may and, of course, there is no problem of freezing as there is with water. When the building heating system is warm air, it is relatively easy to blow the air over the heated rocks. Although this system is a relatively simple one, it requires more power to move the air (as compared with water) and also more heat transfer surface in the collector.

The most common substance for a storage system is water. It is readily available and is a much better heat transfer agent than air. Its specific heat is high. Since it is not to be used for domestic consumption, additives can be used to inhibit corrosion and also to prevent freezing. Since the amount of heat to be stored may be in the order of several million Btu, much water is required for heat storage, perhaps 500 to 1,000 cubic feet. The storage tank can be a very large cylindrical drum or a rectangular concrete tank. If the building heating system is hot water, water from the storage tank can be circulated directly to the radiators. For a warm air heating system, a heat exchanger may be used in which the warm water in the storage tank is used to heat the air.

A considerable amount of research has been done on using the heat of fusion† of various substances for the heat storage system. Substances absorb

---

* See Appendix.

† The heat of fusion is the heat given up by a liquid when it solidifies.

heat as they melt and give up this heat as they solidify. Various salts experience a phase change (from solid to liquid and vice versa) at temperatures close to those of the heat transfer fluid, with a large amount of heat being associated with the phase change. Such organic substances as certain paraffins also offer some promise as heat storage materials. In general, the volume required by these phase change materials is substantially less than that required by either rocks or water. However, the costs of some of these materials must be reduced significantly if they are to gain wide acceptance as heat storage materials. Another problem associated with the use of these materials is that of stratification. As a liquid solidifies, the crystals of the solid tend to collect together in ever growing masses, thus making it difficult to transfer heat from the remaining liquid. It is difficult to freeze a large tank of a liquid since the freezing will occur at the outer portion of the tank where the heat removal is taking place. To continue the freezing process, heat must be transferred from the unfrozen liquid within the tank to its outside surface. It is possible to overcome this problem by using a large number of small tanks rather than one large one, but the cost of the storage system becomes materially higher when this is done.

A simple solar space heating system is shown schematically in Figure 12-3. There is a wide variation in the cost of solar heating systems, even for a given size residence, as can be judged by the number of variables which have been discussed. In certain instances, some home owners have collected materials, in some cases second-hand materials, and put them together themselves to make satisfactory solar space heating systems. Disregarding the value of the labor used, these systems are economically attractive.

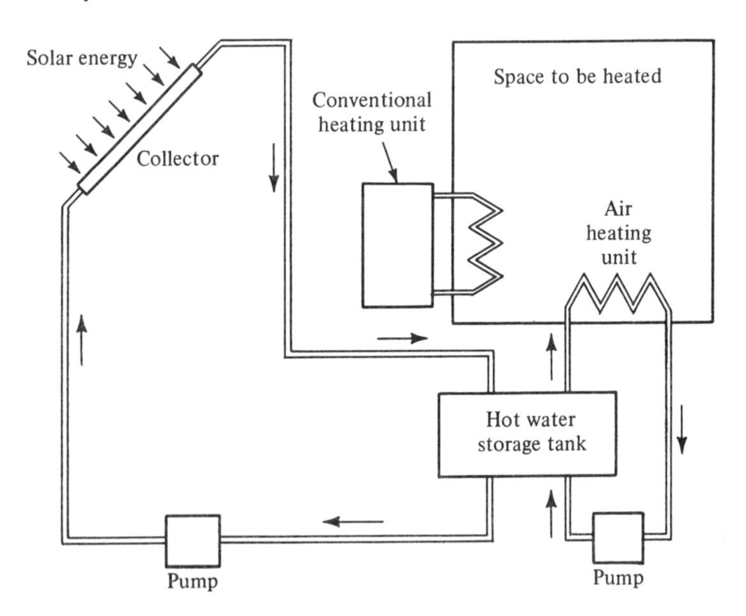

Figure 12-3. Solar space heating (hot water)

The present cost of a completely installed solar heating system, using copper tubing, may range from $10.00 to $20.00 per square foot of collector area. Assuming a collector area of 750 square feet (for a 1,500 square foot area house), the minimum cost of the complete solar heating system will range from $7,500 to $15,000. This size solar heating system will supply the heat needed most of the time. However, for cold, dark days, reliance must be placed on a conventional heating system. This system must be as large as that to be installed without the solar heating system since it must take care of the heating needs when the demands are the highest. Thus the solar heating system does not supplement the conventional system but is an addition to it.

Assume that the conventional heating system uses oil. Depending on the location, the use of a solar heating system could reduce the annual heating bill from $1,000 to $200 or by $800 per year. Assume that the interest rate on the money required to install the solar heating system is 12 percent. Then the annual interest payments on the cheapest solar system will be $900 per year—greater than the savings in fuel. In addition, money must be set aside to repay the original cost plus taxes and insurance on the solar system. Certainly, it will not be desirable to install a solar heating system under these conditions. However, other conditions may make it desirable. Some of these conditions are

1. The labor for the installation is supplied by the owner.
2. Second hand materials are used in parts of the system.
3. The money required to install the system is made available at very low interest rates.
4. Tax credits are given for the installation of the system.
5. The system is exempt from property tax.

At present, it is cheaper to heat with natural gas than with oil. Unless some of the five favorable conditions listed above exist, it is not desirable to replace natural gas heating with solar heating. This may not be true when the price of natural gas is deregulated.

There is considerable variation in the cost of electrical energy throughout the country. In some regions, the cost to residential customers is at least 10 cents per kilowatt hour. Under these conditions, solar heating may be economically desirable. Since heat pumps* deliver much more heat per given electrical input, it may not be desirable to substitute solar heating for heat pumps, particularly in those regions where electricity is relatively cheap.

A considerable effort is being expended to use the solar heating system to produce air conditioning in the summer. One way in which this may be done is to use an absorption system to produce the refrigeration needed to cool the building in the summer. In this system, a liquid absorbs the refrigerating vapor leaving the evaporator and carries it to a generator. In the generator, solar energy boils off the refrigerating vapor which then passes to the condenser.

---

* See the Appendix for a discussion of heat pumps.

After condensing in the condenser, the refrigerant passes to the evaporator to produce a refrigerating effect as it boils. See Figure 12-4. The absorption system, a practical reality for many years in other refrigeration applications, requires relatively high temperatures (perhaps 170°F or higher). Although these temperatures can be attained by using high performance, expensive collectors, the lower collector efficiency and added costs make this use of the absorption system questionable.

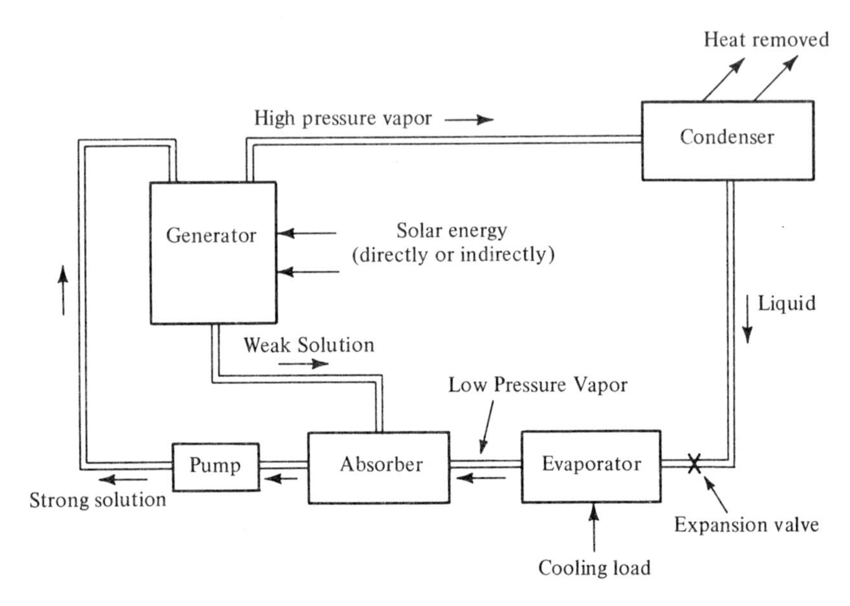

Figure 12-4. Elements of absorption system for solar air conditioning

A second method of using the solar heating system for air conditioning involves the use of a heat pump. In the daytime, in the summer, the heat pump would remove heat from the house and store it in the heat storage system. During the daytime, the solar collector would be covered. At night this cover as well as the cover glass would be removed, with the warm fluid being pumped into it from the heat storage system. The collector would cool this fluid by convection and by radiation into space.

In winter, the collector would operate in the normal manner. The heat pump would remove heat from the collector fluid and deliver it at the desired temperature to the heat storage system. Under these conditions, the collector could be operated at a temperature close to the ambient temperature. Hence, even with no protective covers, the heat loss from the collector should be low. By omitting the covers on the collector, its cost could be reduced significantly, perhaps enough to more than offset the cost of operating the heat pump in the winter. It has been suggested that freon be used as the collector fluid. If this be done, this same fluid would be used in the heat pump, thus eliminating the need for a heat exchanger. More work is required to establish the practicality of this method.

It is difficult to predict the future of solar space heating. However, it does seem possible that there will be some reduction in the cost of the system, particularly in the cost of the collector. That cost may be reduced by improvements in its design and by the substitution of less expensive materials, particularly plastics, for the present materials. Mass production should also help to reduce costs. However, if in an effort to reduce the initial cost, materials are used that will deteriorate rather rapidly, the overall economic gain may be little or nothing. Some enthusiasts predict that it is possible to reduce the initial costs of the collectors themselves to as low as $1.50 per square foot. This low cost is very questionable but if attained for a long-life collector, the solar heating system will become economically sound for most space heating applications. However, most authorities feel that it will be difficult to reduce the cost of a long-life collector to much below $4 to $5 per square foot.

It was demonstrated at least thirty years ago that homes can be heated by solar energy. Today, hundreds of homes are being heated in this manner and perhaps thousands more are being planned. A few buildings, particularly school buildings and governmental office buildings, are solar heated. The federal government as well as some state governments are encouraging further use of solar heating by erecting demonstration homes, by providing low interest loans, and by offering tax credits.

The rapidity of the growth of solar space heating is related to the following factors:

a. The rapid increase in costs of fossil and nuclear fuels, together with shortages of these fuels, will have a tendency to promote the growth of solar heating systems.
b. As experience and engineering know-how is gained and as the solar equipment becomes mass-produced, there should be a large drop in costs.
c. Incorporation of a heat pump or an absorption refrigeration system into a solar energy system will make it more useful throughout the year and should lower the cost of each unit of energy handled.
d. Extensive financial aid by the federal and state governments may make it economically desirable to install solar heating systems.

At present, residential and commercial space heating together require approximately 18 percent of our total energy use. Thus, the potential of solar space heating for saving energy is large. However, even when a million homes become solar heated, the reduction in our energy demands from conventional sources will be less than half a percent. This does not mean that we should not continue to push for solar heating. Rather, we should push harder for it as it becomes economically sound to do so. However, we must realize that it will be several decades before solar space heating will have a significant effect in relieving our demands on other energy sources.

12.3 PASSIVE SOLAR HEATING

Since the solar heating of buildings as just discussed is expensive and probably will remain that way, a considerable amount of effort is now being devoted to passive solar heating. In passive solar heating there is no collector as such. The building is so designed that solar energy enters the building directly in winter. This may be done by placing a large amount of glass area on that side of the building facing the sun. This glass area can be closed off by drapes or curtains to minimize heat losses at night. A sufficient roof overhang is designed to prevent solar energy entering the building during the summer. One way to create passive solar heating in tall buildings was demonstrated in Atlanta, Georgia. The building resembles a pyramid with its top cut off and inverted. Thus each story projects beyond the story beneath it to prevent solar rays from entering in the summer but allowing them to enter in winter.

Various schemes are used in passive solar heating to store energy for nighttime heating and for warming rooms of the building which do not receive solar rays directly. In general, interior walls which receive solar rays are made of brick or concrete which can both transmit energy to other rooms and also store some of it for nighttime heating. In some cases, these walls are designed so that natural circulation of air, which is warmed by the walls, will heat other portions of the building.

Many buildings designed for passive solar heating apparently do not cost much more initially than do conventional buildings. Thus there is a material savings in the overall cost of heating. Although it is very difficult to maintain uniform temperatures with passive solar heating, the savings in heating costs make it increasingly attractive.

12.4 DIRECT PRODUCTION OF ELECTRICAL ENERGY

Various methods are being considered for the direct production of electrical power from solar energy. One of these methods involves the thermoelectric generator. When a bar of either a conducting or a semi-conducting material is heated at one end, a voltage difference is set up between the two ends, thus offering the potential for delivering electrical energy. Solar energy can be used to produce this heating effect. However, the efficiency of such a thermoelectric device is low, even when mirrors are used to focus the sunlight to produce high temperatures. Hence, it is questionable whether solar heated thermoelectric devices will be used extensively.

It has been demonstrated in space vehicles that solar (or photovoltaic) cells can produce electrical energy directly from solar radiation. However, the present costs of solar cells are extremely high, perhaps $100 to $300 per watt of rated capacity. The basic theory of solar cells is very involved since it is based on solid-state physics, quantum mechanics, and statistical thermodynamics.

For our purposes here, consider that a solar cell consists of two semi-conductors, one a p-type and the other an n-type brought together to form a junction. The n-type of a semi-conductor (SC) is doped during its formation with an additive so that when acted upon by heat or by solar energy, some of the electrons will be freed from their atoms, leaving holes in the atoms. The number of electrons freed will exceed the number of holes produced. Conversely, the p-type SC is so constituted that the number of holes produced exceeds the number of electrons which are freed. When sunlight strikes the junction of the two types of semi-conductors, a voltage difference is created. Because of this voltage difference, electrons, and hence electrical energy can be diverted to an external load.

Solar energy may be thought to be composed of very small parcels of energy known as photons. The photons travel as waves, their energies being inversely proportional to the length of the waves. A photon possesses a precise amount of energy. Upon striking an object, it gives up all of this energy and ceases to exist. Depending on the nature of the semi-conductor a specific amount of energy, known as the gap energy, is required to free an electron, the remaining energy being used to produce a heating effect. Thus, even for the best of solar cell materials, such as silicon and gallium arsenide, only 40 to 45 percent of the solar energy received can be used to free electrons, assuming that every particle of solar energy strikes electrons at the junction and that none of the electron energy thus produced is lost. Only a part of the energy imparted to the electron is deliverable. Hence the maximum possible efficiency of solar cells made of other materials, such as cadmium sulfide, is much lower, perhaps around 20 to 35 percent. Efficiencies as high as 20 percent have been obtained in the laboratory for silicon cells but outside of the laboratory efficiencies normally run from 10 to 12 percent for silicon cells and much less for the cadmium sulfide cells.

At present much of the effort in developing solar cells has been directed to the silicon cells, because of their relatively high efficiency and because of the abundance of silicon. Although silicon in the impure state is quite inexpensive, the very pure silicon required for solar cells is extremely expensive. Since solar energy must penetrate one of the two semi-conductor layers to reach the junction, one of them must be made very thin. In the past, this has been accomplished by growing large crystals of silicon and then sawing off extremely thin layers. The initial cost of the silicon, obtaining the very thin layers, and putting together the silicon cells all contributed to the high cost. In 1974, it was announced that the Tyco Laboratories had succeeded in growing fine continuous ribbons of crystalline silicon which could be used directly in solar cells. If this method can be perfected, it appears that there can be a very large reduction in the costs of solar cells. It has been stated that, hopefully it may be possible to reduce the costs of silicon cells from perhaps $400,000 per kW to as low as $1,000 to $1,200 per kW, thus making them competitive with present electric power generating stations. Three factors should be considered at this point:

1. What evidence is there that the costs of solar cells can be reduced to perhaps 5 percent of their present costs?
2. Even if the cost of solar cell per rated kW could be reduced to a cost similar to that of a conventional plant, it must be recognized that the solar cell will produce its rated output only at the middle of the day and that its average output in a 24 hour period is only about fifteen percent of its rated output. This means that solar cells having a rating of four kW will be required to produce the same average output of a conventional plant rated at one kW.
3. Because of low solar intensity in part of the day and no solar energy at night, a large electrical energy storage system must be installed to provide electrical energy when needed. This will add materially to the cost of the total system.

Because of the extremely high costs of solar cells, efforts are being made to use them in connection with focusing collectors. Instead of covering a certain area with solar cells, the solar energy striking the given area would be collected and focused on a small solar cell array. The cost of the focusing collectors and of the small solar cell array together will be much less than that of a larger solar cell array covering the entire area. Since the efficiency of solar cells decreases markedly with an increase in operating temperature, it will be necessary to provide sufficient cooling to maintain the smaller solar cell array at a normal operating temperature, thus attaining normal efficiencies.

Much of the solar energy received by the earth's atmosphere never reaches the earth's surface. In addition, throughout a 24 hour period, any location on the surface receives no solar energy part of the time and reduced amounts much of the rest of the time. To use the energy received by our atmosphere, Dr. Peter E. Glaser of Arthur D. Little, Inc., has proposed to place a five-mile by five-mile collector in synchronous orbin 22,300 miles above the earth. This collector would be covered with silicon cells. Assuming that an efficiency of 20 percent can be obtained for these cells, Dr. Glaser has estimated that 10,000,000 kW of electrical energy can be produced. This electrical energy would be transmitted to a mile square microwave antenna on the space platform and then beamed to a six-mile by six-mile receiving antenna located on the earth. This concept is now under study by other groups, partially funded by government agencies. It is recognized that the intensity of solar energy is much greater at this high altitude. Furthermore the sun will shine 24 hours a day on the collector. Thus the potential for power generation per unit area is many times that for collectors on earth. It should be recognized, however, that the translation of such a design from conceptual stage to a practical reality is a tremendous undertaking. It is possible that if this concept can be demonstrated to be economically feasible, a very large percentage of our electrical power needs could be met by this means. However, until a prototype has been made and operated no prediction can be made as to whether any power will be ever generated by this method.

The storage of electrical energy to meet the needs when the sun is not shining is a difficult problem. Storage batteries can be used for this purpose but the present types of batteries are very bulky and expensive if sizable amounts of electrical energy are to be stored. It has been proposed that the excess electrical energy produced in the daytime be used to produce hydrogen by electrolysis of water, with the hydrogen being burned to produce electrical energy when needed. The use of such a process should be questioned since its overall efficiency is perhaps about 20 percent.

Another method for providing electrical energy at times when solar cells cannot produce has been suggested by Dr. Karl W. Boer of the University of Delaware. A solar house has been operated there under his direction for some time. Part of the roof of the house is covered with cadmium sulfide solar cells. The heat removed to keep the solar cells cool plus the heat from a solar collector mounted on another part of the roof of the house is used to heat the house. Dr. Boer proposes to feed the excess electrial energy generated in the daytime into the local power system and to draw power from that system at night. Not only would such an arragement be advantageous to the home owner but it would be helpful to the power company, assuming that many homes were generating electrical energy from solar energy. Because electrical energy would be delivered to the power system when the demands on it were the highest, this would help the power company to meet its maximum demands without installing additional generating units. If it becomes possible to greatly reduce the initial costs of solar cells, this method of electrical energy exchange with local power companies has merit.

12.5 INDIRECT PRODUCTION OF ELECTRICAL ENERGY

From the discussion in Section 12-4, it appears that it will be several decades, if ever, before large amounts of electrical energy will be produced by the use of solar cells. Because of the anticipated large and growing demand for electrical energy, several groups are working on schemes to use solar energy to generate steam, with the steam to be used in steam turbo-generators for electric power production. As was discussed in Chapters 1 and 3, it is essential to generate steam at high temperatures and pressures if high thermal efficiencies are to be attained. Steam under these conditions can be generated only by focusing and concentrating the solar energy.

One concept for concentrating and using solar energy for power production was developed by a husband-wife team, the Meinels, at the University of Arizona. The Meinels* proposed to cover some 3,000 square miles of desert area in southern Arizona, southern Nevada, and eastern California with solar collectors. In many ways, this is an ideal location since there are very few cloudy days and, at present, the land is more or less worthless for other purposes. They proposed special cylindrical collectors through which a liquid

* Aden B. Meinel and Marjorie P. Meinel, "Physics Looks at Solar Energy," *Physics Today*, February 1972, p. 44.

metal would be circulated. Temperatures as high as 1,000°F were projected. The heated metal would heat up an eutetic material in large storage tanks to store energy for night operation. Steam would be produced to be used in turbo-generators. It was estimated that 1,000,000 megawatts of electric power could be produced, sufficient to meet American energy needs for years to come. This scheme is ingenious but involves many problems. Also, until a prototype plant has been constructed the economic feasibility of this proposal cannot be determined.

Much of the effort to develop solar power plants is now being devoted to the "power tower" concept. The power tower concept uses a large number of collecting mirrors to collect solar energy and focus it on to the top of a tower where it is used to produce steam, directly or indirectly, for use in a turbo-generator. In 1975 the Energy Research and Development Administration (ERDA) awarded contracts totaling 8 million dollars to four industrial teams, Honeywell Systems and Research Center, McDonnell Douglas Astronautics Company, Martin Marietta Aerospace Company, and Boeing Company, to develop plans for a 10 megawatt pilot solar power plant using this concept. As an example of the size of these plants, McDonnell Douglas proposes to use 2,470 tracking mirrors in a field 1,800 by 2,000 feet and to deliver the solar energy to a tower 312 feet high. As a result of this study, a 10 MW solar power plant is now being constructed. The proponents of this method of producing electric power are quite enthusiastic. One projection foresees 40,000 megawatts of electric power produced from this type of plant by the year 2000. It is not possible to evaluate the soundness of this prediction until operating information becomes available.

Another development in solar power production is the solar pond. Solar rays penetrate fresh, clear water ponds to a considerable depth, warming the water. The warm water, being less dense, rises towards the surface where it is cooled off. Thus the main body of the water does not attain a temperature greatly in excess of the ambient. However, when salt is dissolved in the water in the bottom of the pond, this salt water, being dense, will not rise even where heated to relatively high temperatures by the solar rays. Thus this heated salt water can be used to vaporize a liquid, producing a vapor to be used in a turbine for power production.

This concept has been used at Ein Bokek, near the southern shore of the Dead Sea in Israel. A pond having a depth of 8 feet and an area of 70,000 square feet is used. Salt water at the bottom of the pond reaches a temperature of around 175°F. This hot water is pumped to a heat exchanger where it vaporizes a liquid, such as a Freon. The vapor is fed into a turbo-generator where it produces 150 kW. The vapor then passes into a condenser where it is condensed by water drawn from the surface of the pond. Based on the experience gained from this pond, Israeli officials are planning to build a 5,000 kW solar pond generating unit. In this larger pond, criss-cross baffles must be used to prevent winds from churning up the water. If this pond is successful, there are plans to install many more of these ponds around the Dead Sea. Ultimately such ponds could supply a third of Israel's electrical needs by the end of this century.

The results from this first solar pond have generated enthusiasm for solar ponds in the United States. A 5,000 kW solar pond power plant is being planned at the Salton Sea in the Imperial Valley in California. With the aid of government funding, investigations are being made into developing solar pond power in such locations as San Francisco Bay and Great Salt Lake in Utah. It is very difficult to predict the potential of solar pond power but some proponents estimate that ultimately 12 percent of our power could be produced by this method.

12.6 SUMMARY ON SOLAR ENERGY

The growing scarcity of conventional energy supplies and their rapidly increasing costs make it mandatory that we seek other sources. One obvious source is solar energy. The earth receives sufficient solar energy to meet present energy needs many times over. However, sunlight is so diffuse that it is very difficult and expensive to capture.

An increasing effort is being expended to use solar energy for heating and for electric power production. For the latter both solar cells and solar steam power plants are being considered. Some efforts have been devotd to using solar energy for such diverse purposes as crop drying and metallurgical furnaces. Much more effort is being devoted to the use of solar energy for heating water and buildings. Solar heating of domestic hot water has been demonstrated to be economically sound in some regions of the country. It appears that as mass production lowers the cost of solar equipment and as the price of conventional fuels increase, there will be a large increase in the use of solar hot water heaters. Desirable as this is, however, this use of solar energy will not cause much of a decrease in our demands for other types of energy for many years to come.

Although there is much interest in solar heating of buildings, at present this method of heating has not been demonstrated to be economically sound. However, in most cases, if improvements can be made in the design, if construction costs can be reduced significantly by mass production, and if government provides low interest loans and tax relief, the number of buildings heated by solar energy will increase significantly. This number may increase if it also becomes feasible to use the solar heating system to provid summer air conditioning. However, it is difficult to imagine that the majority of homes and other buildings will become solar heated and perhaps cooled in the next one or two decades. Hence, this use of solar energy will not bring about a significant reduction in our demands of energy from other sources in the near future.

It has been demonstrated, particularly in conjunction with spacecraft operation, that solar cells can produce electrical energy. However, solar cells are presently so expensive that there is little chance of their widespread use unless their costs can be reduced by a factor of at least 20 to 40. Some significant reductions probably will be made by improvements in design, in methods of construction, and by the use of focusing collectors. It does not now appear that the necessary reductions can be made in the next decade or two to justify extensive use of solar cells.

The potential for generating large amounts of electrical energy in solar-steam plants in the hot, dry regions of the country is high. However, many problems such as waste heat disposal from the power plant need to be solved before the feasibility of this system can be established. Since operating data is not available for even very small plants it is not possible to make accurate estimates of the cost of power from this type of plant. It is very difficult to envision sizable amounts of power being produced by solar-steam power plants in the next two decades.

SUPPLEMENT A

ILLUSTRATIVE EXAMPLES

Example 12-1. A solar cell "farm" is to be used to produce 100,000 kW of electrical power on the average, per day. The farm is to be located in a region near Omaha, Nebraska. Assume that the average efficiency of the solar cell is 10 percent. Determine the land area required in acres and square miles.

Solution. From Table 12-1, the average solar insolation in Omaha is 1,351 Btu per sq. ft. per day. Then the average electrical output is

$$0.1 \times 1{,}351 = 135.1 \text{ Btu per day per sq. ft., or}$$

$$\frac{135.1}{24} = 5.63 \text{ Btu per hour per sq. ft., average}$$

Then the number of square feet required equals

$$\frac{100{,}000 \times 3{,}413}{5.63} = 66{,}200{,}000 \text{ sq. ft., or}$$

$$\frac{66{,}200{,}000}{43{,}560} = 1{,}520 \text{ acres, or} \qquad \text{answer}$$

$$\frac{1{,}520}{640} = 2.38 \text{ sq. miles} \qquad \text{answer}$$

Example 12-2. A solar collector for house heating is located in a region near Omaha, Nebraska. The house loses on the average 30,000 Btu per hour in December. The collector has an area of 900 square feet and has an efficiency of 55 percent. What percentage of the heat loss form the house will, on the average, be supplied by solar energy in December?

Solution. From Table 12-1, the average solar insolation in December is 596 Btu per sq. ft. per day. The collector then will supply

$$596 \times 0.55 \times 900 = 295{,}000 \text{ Btu per day}$$

The percentage of the needs is

$$\frac{295{,}000 \times 100}{30{,}000 \times 24} = 40.9 \text{ percent} \qquad \text{answer}$$

Example 12-3. A house is heated by the use of oil at a cost of $900 per year. It is estimated that 80 percent of this cost can be saved by the use of a solar heating system. Assuming total fixed charges to be 15 percent and neglecting all operating costs, calculate the maximum economic cost of a solar heating system.

Solution. The savings in fuel is

$$0.8(900) = \$720 \text{ per year}$$

Maximum initial cost of solar heating system is

$$\frac{720}{0.15} = \$4{,}800 \qquad \text{answer}$$

SUPPLEMENT B

PROBLEMS

12-1. Assume that the solar cells in Example 12-1 have an average output of 15 percent of their rating. Assume also that the cost per kW of rating is $30,000 per kW. Determine the cost of the solar cell plant, neglecting land costs.

12-2. Assume for Problem 12-1 that the total fixed charges are 15 percent. Determine the costs of power delivered by the plant, in cents per kWh, assuming no operating costs.

12-3. a. Same as Example 12-2 except for February. Assume the same heat loss from the house.
b. Make a comment on the percentage of the building heat, on the average, which can be supplied by a solar heating system.

12-4. Same as Example 12-3 except that because of a low interest government loan and tax credit, the fixed costs are 7 percent.

12-5. The solar heating system in Example 12-2 replaces electrical heat. If electrical energy costs 5.2 cents/kWh, determine the cost of the electrical energy saved per month.

12-6. Discuss why solar energy is used more extensively for water heating than for space heating.

12-7. The collectors of a solar-steam power system absorb 88 percent of the solar energy they receive. Assume that 32 percent of the energy absorbed is delivered as electrical energy and 98 percent of the remainder is rejected in the condenser. Determine the amount of heat rejected per hour in the condenser when the power output is 1,000,000 kW.

12-8. If the temperature rise of the circulating water passing through the condenser in Problem 12-7 is 15°F, determine the tons of water required per second.

12-9. Discuss the desirability of extensive government funding for solar research and development. Should some areas be stressed in preference to other areas?

12-10. It has been stated in various parts of this chapter that it is desirable to attain high efficiency in collecting and using solar energy. Why should this be necessary since solar energy is free?

12-11. A home in Burlington, Vermont, uses 290 gallons of oil in December. Assume that the oil has a heating value of 140,000 Btu per gallon and is burned with an efficiency of 70 percent. It is proposed to use solar collectors to provide all of the heating. The overall efficiency of the solar collectors is 48 percent. Determine the collector area required.

12-12. Same as Problem 12-11 except in Atlanta, Georgia, where the amount of oil used in December is 195 gallons.

12-13. Using the results from Problem 12-11 and Problem 12-12 comment on the desirability of completely heating a home by solar energy.

13

Wind Power

13.1 INTRODUCTION

Those who have witnessed the destructive power of hurricanes and tornados recognize the very large potential for power production from the winds. It has been estimated that approximately two percent of the solar energy received by the earth is converted into wind energy. This means that wind energy produced in the continental United States is approximately 13 times our total energy use in 1979. What are the possibilities of using this energy? Perhaps 30 percent of the wind energy is developed in the lower 3,200 feet of our atmosphere, but it is not feasible to try to use much of this wind energy, particularly at elevations much above 1,000 feet. Fortunately, as energy is removed from the winds at the lower elevations by any means, the higher velocity winds at the upper levels tend to drag along the air at the levels, thus partially restoring the wind velocity at the lower levels. Of course, it is extremely difficult to estimate the total economically recoverable energy of our winds when the wind system is such a dynamic one. However, a research team at the University of Maryland attempted to do this. They estimated that it will be possible to develop wind power equivalent to our total electric use in 1972.

Windmills apparently were first developed in ancient Persia. Their use spread gradually throughout the Islamic world and then into China. Windmills first appeared in Europe in the 11th century. By the 17th century, the Netherlands became industrialized by extensive use of wind power in ships and on land, particularly for pumping. Windmills were used extensively in this country in the first part of the 20th century to pump water and to run sawmills. Thousands of windmills were equipped with small electrical generators (generally a fraction of one kW) to provide some electrical energy in rural areas where no electrical energy was available. At one time, these small windmill-electric generators could be purchased from a mail-order catalog. However, with the spread of rural electrification, which supplied rural customers with relatively low cost electricity, it became economically unsound for the rural inhabitants to generate their own electricity from the wind.

13.2 LARGER SCALE DEVELOPMENT OF WIND POWER

In the last 50 to 60 years, several large windmills have been constructed in the U.S.S.R. and in Great Britain. However, they were not commercially successful. In 1941, a 1,250 kW wind turbo-power plant was built on top of Grandpa's Knob near Rutland, Vermont. It consisted of a two-bladed rotor, having a tip diameter of 175 feet and mounted on a 110 foot tower. In October of that year, it was phased into the lines of the local power company. It operated until February 1943 when the failure of a main bearing caused a shutdown. Because of wartime shortages of metal, a new bearing was not installed until early in March 1945. Later that month, a structural failure caused the loss of one of its blades. Although the technical feasibility of using wind energy for electric power production was demonstrated, it appeared at that time that such a system was not economically sound. Hence no effort was made to repair this wind power plant.

The abundance and low costs of fossil fuels and the extension of power lines into the rural areas left little incentive until recently for developing wind power. Now, however, with the increasing costs and also shortage of fossil fuel, much more attention is being paid to the possibility of developing wind power on a larger scale. A few companies in this country and abroad are manufacturing wind power generating systems of several kilowatt capacity for use in remote areas. A few years ago what was at that time the Energy Research and Development Administration (ERDA) in cooperation with the National Aeronautics and Space Administration (NASA) built a 100 kW windmill mounted on a 100-foot tall tower in Sandusky, Ohio. It has a two-bladed rotor, with a tip diameter of 125 feet. It produces its rated power at a blade rotation of 40 rpm with a wind speed of 18 miles per hour. The blade speed is maintained constant at other wind speeds by feathering* the blades. The blades become full-feathered at wind speeds less than 8 miles per hour and also above 60 miles per hour. Under these conditions, no power is produced. The rotor is connected to a 1,800 rpm, 60 cycle generator.

The Department of Energy (DOE), which supplanted ERDA, constructed a 2,000 kW windmill power plant in the late 1970's near Boone, North Carolina. It produced power satisfactorily but was so noisy that nearby inhabitants objected. To reduce the noise the speed of the rotor was reduced in 1980. This reduction in speed minimized the noise problem but also reduced the maximum power output to 1,500 kW.

Efforts are being made both in this country and in Europe to build larger wind power plants. Messerschmidt-Bolkow-Blohm (MBB), Germany's leading

* Feathering means changing the angle of the blades, hence changing the resistance they offer to the wind. When the blades are fully feathered, the blades are so turned that they offer no resistance to the wind.

aerospace company, plans to build a 5,000 kW windmill in the early 1980's. This windmill will be unique in that it will be single bladed, counterweighted. The blade, 238 feet long and weighing 26 tons, will be mounted on a 394-foot tower. It will operate at 17 rpm. Before building this large unit, MBB is in the process of building a smaller unit to obtain operating data. This smaller unit, rated at 370 kW, has a 79-foot blade mounted on a 164-foot tower. It will be operated at 44 rpm. In this country, the Department of Energy is planning to erect three windmills at Goodnoe Hills, Washington. These windmills, built by Boeing, are two-bladed, measuring 300 feet from tip to tip. They are to be mounted on 200-foot towers and are rated at 2,500 kW. The Department of the Interior, working jointly with NASA Lewis Research Center, is planning windmills of 4,000 kW rating. The blades are to be 225 feet long and will be mounted on a 262-foot tower. They will be erected in Medicine Bow, Wyoming, and will be constructed by Hamilton Standard Corporation. If these windmills produce as predicted, it is possible that 50 of them ultimately will be erected.

13.3 THE FUTURE OF WIND POWER

Several factors must be considered in trying to predict the future use of wind power:

1. The energy of the wind entering a given size windmill varies as the cube of the wind speed. A windmill rated at 27 kW at a wind speed of 24 miles per hour will produce only about 1 kW with a wind speed of 8 miles per hour. Hence, windmills are feasible only in those regions where wind velocities are high.
2. Because of the high costs of wind power plants, it is essential that they deliver power a large portion of the time. Therefore they should be located in areas where the wind is relatively steady.
3. If the electrical energy generated is to be used in AC, 60 cycle devices and the electric generator is directly connected to the windmill, the windmill rotor must be operated at constant speed. An alternative to this is to generate DC current and then use an alternator to produce AC current, 60 cycle. Both of these methods are expensive.
4. An electric power storage system of some sort must be provided to furnish power at times of low wind velocities. One possibility is to integrate the wind power generating system into a local power system. This should be satisfactory, provided that the output of the wind power system is relatively low compared with that of the power grid to which it might be tied. For storage of small amounts of power, storage batteries may be used. However, present batteries are expensive. Other suggestions for storage include the use of large flywheels, the generation and storage of compressed air and the production of hydrogen. The economic feasibility of these various storage devices has not been established.

5. Various proposals have been made and are being investigated as to the optimum configuration of the wind power unit, involving the placement of the rotor (horizontal or vertical), the nature of the blading system, and the optimum number of blades. Also in progress is an investigation of the total wind power system.

The average reported cost of a 6,000 watt wind power generating unit was said to be around $6,000 in 1979. This price is for the unit alone and does not include the tower, the batteries, the controls or the inverter. These would run the total cost up to approximately $13,000. At 12 percent interest rate, the interest on this money will be $1,560 a year. Even in the best location, probably the average power output will be about 20 percent of the rated or 10,500 kW hours per year. This means that the interest cost alone will be about 15 cents a kW hr. In addition, the original $13,000 must be paid back and, unless they are forgiven, taxes are to be paid plus insurance. Thus the cost of electricity will be close to 25 cents a kilowatt hour.

These figures show that generation of electrical energy by the use of small windmills is out of the question except for those areas which are remote from the power company lines. For such areas it may be desirable to use wind power generators. However, the total amount of power which will be generated in the remote areas is only a very small fraction of the total electric power requirements for the country as a whole.

The Great Plains region of our country is favorable to wind power development. Winds are relatively steady and velocities are relatively high. The potential for wind power development in this area is shown in Table 13-1. This table is based on data presented in *Energy Alternatives: A Comparative Analysis*, prepared for several government agencies by The Science and Public Policy Program, University of Oklahoma, May 1975.

Table 13-1. Annual Windmill Power Electrical Output in Great Plains Region

Windmill Diameter		Installed Capacity	Energy Output
Meters	Feet	kW	kWh per yr.
10	32.8	31	49,000
20	65.6	126	196,000
30	98.4	282	442,000
40	131.2	502	785,000
50	164.0	785	1,227,000

Two factors should be noted from this table. First, the rating of the units increases as the square of the diameter. Second, even in this most favored region, the total yearly power output is only about 18 percent of the rated output, compared with approximately 55 percent for a conventional power plant. This means that the installed capacity of the wind power plant must be at least 3 times that of a conventional plant for the same average output. Of course, in less favorable regions, the wind power plant must be much larger for a given average power output.

The potential for wind power development in the Great Plains region has been recognized by several investigators, particularly Prof. William E. Heronemus of the University of Massachusetts. He estimates that 30,000 wind turbine units located in the Great Plains and stretching from North Dakota to Texas could produce 189 million kW.* Heronemus has also investigated the potential of wind power offshore along the coast of the eastern United States. He suggests that floating towers be anchored some distance offshore as far out as the edge of the continental shelf and stretching up and down the Atlantic coast for 1,000 miles. He has considered various configurations for the windmills. Specifically he proposes to use towers ranging in height from 100 to 500 feet. Each tower may have three windmills mounted on it, the windmills ranging in size capacity from 20 to 120 kW. If proper allowances for sufficient distances between the windmills to avoid interference with the air flow are made, Heronemus estimates that it is possible to install windmills offshore to have a total rating of 267,000,000 kW. He has calculated that the yearly output per kW of rating will vary from 3,000 to 4,000 kWh, depending on the particular location of the windmill.

Consideration is being given to using the electrical energy developed offshore in several ways:

1. For those wind power plants closest to the shore, the electrical energy would be delivered to the local power system.
2. Hydrogen could be produced by electrolysis at the windmill site. This hydrogen could be piped, underwater, to shore where it might be substituted for natural gas. It also could be used as an automotive fuel, provided that satisfactory methods can be developed for this use of hydrogen.
3. The electrical energy could be used for nitrogen fixation for the production of fertilizers.

There are several advantages in the use of wind power. It is a renewable source. It is non-polluting. However, the question has been raised as to whether or not the installation of a very large number of windmills rather close together could change weather patterns. It is possible that the installation of a very large number of windmills may interfere with air and sea navigation. Then, too, there are some objections about the appearance of a large number of windmills dotting the landscape. Finally, and most important of all, what is the economic feasibility of this method of power production? Heronemus has made detailed estimates of the cost of producing power from windmills on a large scale basis and has concluded that it is economically sound, particularly with rising costs of conventional energy sources. However, it is very hazard-

* The total installed electric generating capacity in this country at the end of 1980 was 600 million kW and is estimated to be 800 million kW by 1985.

ous to predict costs for large scale units when smaller units have not been produced. Conceivably, he may be correct or he could be greatly in error. Before decisions are made as to the desirability of proceeding with Heronemus' or other similar proposals, much information should be obtained from actual operating units.

13.4 SUMMARY ON WIND POWER

The winds of the earth possess vast quantities of energy, much more than ample to meet our energy needs if they could be harnessed. Furthermore they are a renewable source and are non-polluting. However, except for violent storms, the energy of the winds is diffused and difficult to capture. There is wide variation in the wind velocity at any given point and also much variation in the velocity from location to location. Since the potential power for a given size windmill varies as the cube of the wind velocity, it is difficult to design a machine which can use its rated capacity much of the time.

Wind has been used, on a small scale basis, as a source of power for a long time. This was particularly true in rural areas until the availability of cheap electric power made it economically unwise to continue using wind power. At present, it appears that electricity produced from wind power on a small scale is much too expensive to compete with electrical energy obtained from conventional sources. However, when conventional electric power is unavailable, wind driven electric generators can satisfy a need.

Extensive studies have indicated a vast potential for wind power development both in the Great Plains region and offshore at the Atlantic seaboard. It has been predicted that it will be possible by wind energy use in these regions to meet a large portion of our demands for electrical energy. However, the economic feasibility of this concept has not been established and cannot be established with certainty until cost information becomes available from actual units of the size contemplated for use. Certainly, with improvement in design, with mass production of the units, and with rising energy costs, it may become much more feasible to use the energy of the winds for power production than it is at present.

SUPPLEMENT A

ILLUSTRATIVE EXAMPLES

Example 13-1. A windmill electric generating system having a diameter of 120 feet has an efficiency of 45 percent. Determine its power output for a wind velocity of 20 miles per hour. The air density is 0.0748 pounds per cubic foot.

Solution. The velocity in feet per second equals

$$\frac{20 \times 5{,}280}{3{,}600} = 29.33 \text{ ft. per sec.}$$

The flow area is $\frac{\pi \times 120^2}{4} = 11{,}310$ sq. ft.

The volume rate of flow is $11{,}310 \times 29.33 = 331{,}700$ cfs.

The mass rate of flow $= 331{,}700 \times 0.0748 = 24{,}810$ lb./sec.

The kinetic energy $= \frac{24{,}800 \times 29.33^2}{2 \times 32.17} = 331{,}700$ ft. lb./sec.

The horsepower output $= \frac{331{,}700}{550} \times 0.45 = 271.4$ hp.

The electrical output $= 271.4 \times 0.746 = 202.5$ kW. answer

Example 13-2. Determine the maximum costs per kW rating of a large wind-power electric generating system when the average production of power is 18 percent of the rating. Fixed costs are to be 14 percent. Power is to be produced for 8 cents/kWh.

Solution. The rating, per average kWh produced equals

$$\frac{1}{0.18} = 5.556 \text{ kW}$$

Let x = cost per kW rating. Then the fixed charges, per average kWh produced, equals

$$5.556 \times 0.14x = \$0.7778x \text{ per year, or}$$

$$\frac{0.7778x}{8{,}760} = \$0.0000887x \text{ per kWh}$$

where 8,760 is the number of hours per year.

Then $0.0000887x = 0.08$, $x = \$901.9$. answer

SUPPLEMENT B

PROBLEMS

13-1. You wish to produce your own electric power. Assume that the average monthly load will be 800 kWh. Assume that a wind power system will deliver on the average 17 percent of its rated capacity. This system, completely installed, will cost $3,000 per rated kW. Fixed charges are 14 percent. Determine the cost of power per kWh assuming no operating expenses.

13-2. Suggest specific ways in which the cost of the power produced in Problem 13-1 may be reduced.

13-3. Consider using a gasoline-driven electric generating set for Problem 13-1. A suitable unit can be bought for $1,600. Because of its relative short life, assume the fixed charges to be 25 percent. The generator set produces 5.5 kWh per gallon of gasoline. Gasoline costs $1.30 per gallon. Determine the cost of power in cents per kWh.

13-4. Discuss the desirability of using wind power electrical energy developed offshore to produce hydrogen to be piped ashore for power production.

13-5. A windmill electric generating system is to be designed to produce 2,000 kW in an 18 mile per hour wind. If the efficiency is 46 percent and the air density is 0.0748 pounds per cubic foot, determine the diameter of the windmill.

13-6. Should equal consideration be given to the use of wind power offshore of the Pacific coastal states as is being given to the Atlantic coastal states?

13-7. Same as Example 13-2 except the ratio of the average power produced to the maximum is 25 percent.

13-8. The absolute maximum wind velocity at a particular location is 54 miles per hour. The wind power unit is to produce a maximum of 28 kW. Should the unit be designed to produce the 28 kW with a 54 mile per hour wind?

13-9. A family uses an average of 1,020 kWh per month. The output of a windmill in the area is 12 percent of the rating. Calculate the design rating of the windmill.

14

Geothermal Energy

14.1 INTRODUCTION

It is well known that the interior of the earth is molten. As such it contains vast quantities of energy. Some heat is conducted outward through the earth's crust and is dissipated from the earth's surface. In spite of this heat loss, the temperature within the earth seems to remain substantially constant, apparently being maintained by the decay of radioactive material.

It has been estimated that the total energy content of the earth's crust underlying the 50 states up to a depth of 6½ miles and having a temperature above 300°F is about 13.2 million Quads. This energy is about 170 thousand times the total present annual energy consumption in the United States. If 2 percent of this energy can be used it would be sufficient to meet all of our non-transportational needs for about 2,000 years at the present rate of energy consumption. Hence geothermal energy has a high potential for supplying much of our energy for centuries.

The concept of using geothermal energy is not new. Power has been produced from geothermal steam in Larderello, Italy, since 1904. For many years, geothermal hot water has been used for building heating, particularly in Iceland and in a few locations in this country. The total geothermal energy used to date, however, is very small. Spurred by our growing energy shortages, much attention is now being given to the use of this energy.

The nature as well as the thickness of the earth's crust, which covers the magma (molten rock) is quite variable. The earth's crust has a tendency to be thinner in those regions of relatively recent geological activity. The depths to which it is necessary to go to reach desired temperatures for energy use depend not only on the thickness of the earth's crust but also on its composition and hence its thermal conductivity.

As shown schematically in Figure 14-1, in some regions of the country, ground water penetrates to considerable depths, where it is heated by the hot rocks.

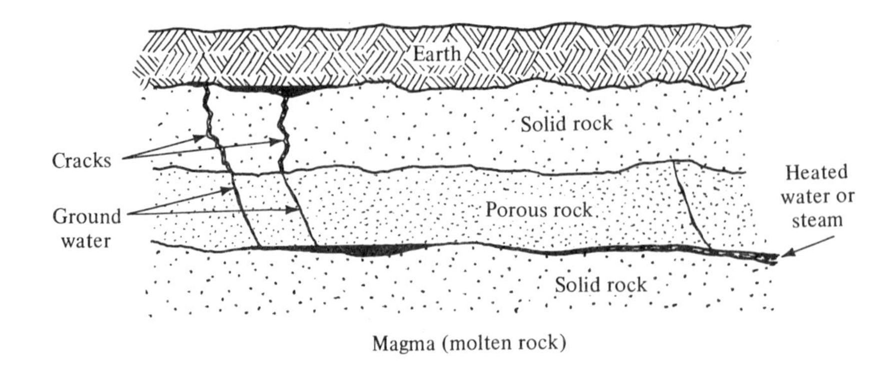

Figure 14-1. Geothermal heating of water

The temperature to which the water is heated in the porous rock is dependent on the rate of flow of water, on the rate of heat flow, and on the restrictions placed on the flow of water after it is heated. In some cases the water may be heated to as high as 400°F to 500°F. Frequently, there are fissures in the rock above the porous rock, allowing the heated water to rise to the surface of the ground. Hot springs are a manifestation of this action. Since the porous rock is at a considerable depth, the water in this rock is under pressure, in some cases several hundred pounds per square inch. Because of this pressure the water may be heated up several hundred degrees without boiling. If, however, the rate of flow of water in a particular segment of the porous rock is relatively low, the water may be heated sufficiently to boil it, thus forming steam.

In many parts of the eastern United States there are indications that geothermal water temperatures may reach as high as 220°F to 250°F at depths not exceeding a mile or two. Although such a temperature is considered too low for power generation, this water can be used in poultry plants, for vegetable and seafood processing, for grain, for timber and tobacco drying as well as in other industries. Water at this temperature—as well as a somewhat lower temperature—has much potential for building heating. The Department of Energy has funded studies on the extent of this geothermal hot water, the temperature of the water, and the depths at which it may be found. These studies are vital to predictions of the potential of geothermal hot water in making a sizeable contribution to the energy resources in the region east of the Rockies.

When temperatures are sufficiently high, geothermal energy can be used for power production. Three possible methods of geothermal power production are

1. Geothermal steam
2. Geothermal hot water
3. Energy from hot rocks (dry rock method)

14.2 GEOTHERMAL STEAM

Geothermal steam, in significant quantities, is produced in only a few regions of the world. In this country the only region in which large quantities of geothermal steam are produced is at the Geysers, located about 90 miles north of San Francisco. For many years, man was intrigued with the large quantity of steam issuing from the ground at the Geysers. An initial attempt was made in 1921 to use this steam but it was not until 1960 that the first successful use of this steam was made. Holes were drilled down to the main body of the steam, capturing it and delivering it to a steam turbine. Since 1960, many more holes were drilled and more turbines were installed. The total capacity of the geothermal steam turbines at the Geysers is approximately 900,000 kW. This is close to 0.15 percent of the total electric generating capacity in this country. Additional units continue to be added. The total potential of this geothermal field is uncertain but it is felt that there is sufficient steam being generated to produce at least 3,000,000 kW and perhaps as much as 4,000,000 kW. It has been predicted that steam will continue to be produced at the desired rate in the Geysers region for at least 50 years. However, there is much uncertainty in this prediction.

Figure 14-2 presents a line diagram of a geothermal steam power plant. After passing through the turbine and producing work, the steam flows into the condenser. Cool water from the cooling tower is injected into the condenser*, thus condensing the steam. The warm water leaving the condenser is piped to the cooling tower, where it is cooled, largely by evaporation. Per pound of steam condensed in the condenser about 0.8 pounds of water are evaporated in the cooling tower. The other 0.2 pounds of water may be reinjected into the earth to help compensate for the steam which was removed.

It may appear that the cost of power produced from geothermal steam should be very low since the steam is free. This is not the case, however. The total initial investment cost is high. It is expensive to drill holes to the required depths to reach the steam sources. (At the Geysers, the holes range from 600 to 9,000 feet deep.) The steam is at a relatively low pressure and temperature—115 psia (pounds per square inch absolute) and 348°F on the average, compared with steam conditions of 2400 to 3500 psia and 1000°F for a modern fossil fuel power plant. Because of the low steam pressure and temperature, the efficiency of the steam plants at the Geysers is very low, being approximately half of that of a modern fossil fuel plant. This means that all of the steam units at the Geysers must be materially larger than those of a conventional fossil fuel plant having the same electrical output.

However, no furnace is required for fuel burning so the overall cost of the geothermal steam power plant at the Geysers is low enough for power to be delivered at a cost significantly below that from a conventional fossil fuel

* This is a direct contact condenser and does not have tubes as does the conventional condenser.

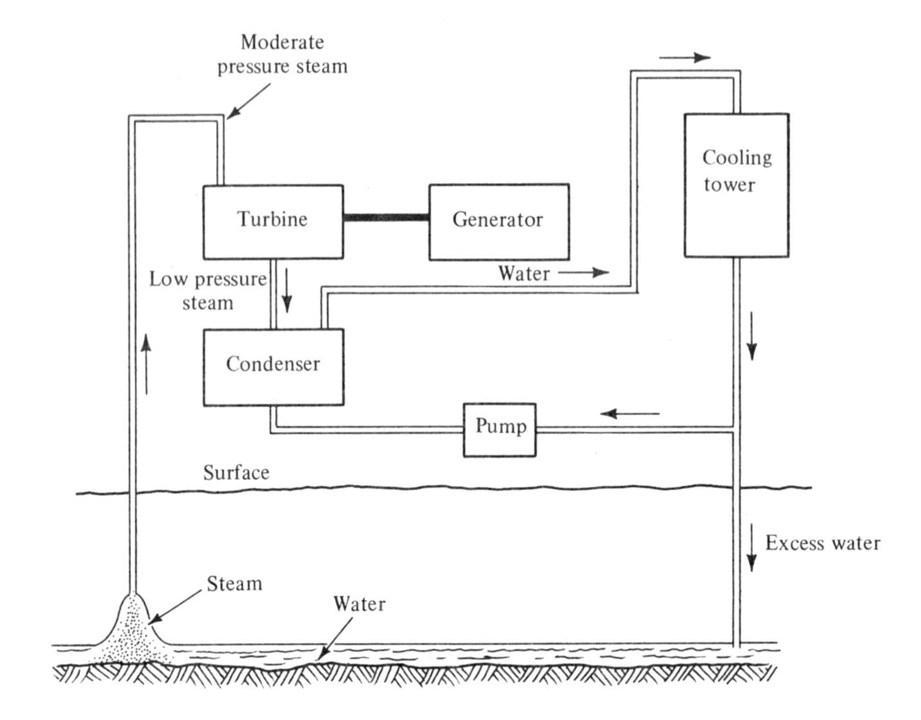

Figure 14-2. Elements of a geothermal steam power plant

plant. It is because of this lower cost of power generation that units are being added at the Geysers at a rate of 110,000 kW per year.

There are some environmental problems connected with the use of geothermal steam. The steam coming from the wells at the Geysers contains about 1 percent of non-condensable gases. The predominate gas is carbon dioxide. Other gases are ammonia, methane, hydrogen sulfide, nitrogen and hydrogen. Because of its unpleasant odor, hydrogen sulfide is very objectionable. At present, there is no completely satisfactory method of removing it from the steam. It has been estimated that the amount of sulfur in the hydrogen sulfide at the Geysers power plants is approximately the same as that discharged by a coal burning power plant burning a low sulfur coal when producing the same electrical output.

14.3 GEOTHERMAL HOT WATER

Although the regions where geothermal steam is found are very limited, there are very many areas where geothermal hot water is quite abundant. These known areas are listed by states in *Geothermal Energy Resources and Research.** Most of these areas in this country lie in the Rocky Mountain and Pacific Coast states. In addition to the known areas, it is believed that a large

* U.S. Cong., Senate, Committee on Interior and Insular Affairs, *Geothermal Energy Resources and Research*. Washington, D.C., hearings, GPO, June 15 and 22, 1972, p. 438.

number of areas exist which presently have not been identified. It will require extensive test borings to determine, even approximately, the true potential of geothermal hot water for power production. It should be noted that many of the areas having potential for geothermal hot water power production lie on federal lands.

There is a very wide range in the characteristics of geothermal hot water. In some of this water, the salinity may be as much as 200,000 ppm (parts per million). This is approximately six times that of normal ocean water. In some cases, the water temperature is too low for power production. In a few cases, water temperatures may be at least as high as 500°F. In some areas, steam is mixed with the hot water. In others, no appreciable amount of steam is present. It is evident, then, that the method to be used in the production of power from geothermal hot water is dependent on the nature of the water.

At this point, it is well to call the attention to several facts relevant to hot water and steam. When the pressure on hot water is reduced, some of it will flash into steam. For instance, if the pressure on water boiling at 400°F is reduced from 247 psia to 150 psia, approximately 5.1 percent of it will flash into steam which can be used in a steam turbine. If the pressure is further reduced to 50 psia, an additional 8.4 percent of the original water will flash into steam.

A pound of steam at 50 psia will produce about three quarters as much work as steam at 150 psia. Furthermore the volume of a pound of steam at 50 psia is approximately 2.8 times the volume of steam at 150 psia. Thus a steam turbine receiving steam at low pressure must handle a very much larger volume of steam for the same output as will be handled by a higher pressure turbine. It should be evident from the above discussion that the temperature of the geothermal water as it is withdrawn from the earth has a marked effect on the nature of the power plant as well as on its size.

When geothermal water contains steam, the steam may be separated and used directly in a steam turbine. (See Figure 14-3.) If the remaining hot water is at a high temperature, some of it may be flashed into steam which is supplied to a low pressure turbine as shown in Figure 14-3. However, when the water temperature is low (below 350°F), only a small amount of steam can be produced by a pressure reduction. Furthermore because of the low pressure, the volume of the steam would be very large, thus requiring a large turbine for a given output. To avoid using a large steam turbine, using the hot water to boil another fluid is proposed, such as isobutane, which boils at a relatively high pressure and hence has a much smaller volume. This vapor is used in an especially designed turbine. (See Figure 14-4.)

When geothermal water containing no steam has a high temperature (above 350°F), it may be flashed into steam as described above. For lower temperature water, it may be used to boil a secondary fluid, as described above.

There are several possible adverse environmental effects associated with the use of geothermal water. The amount of water to be withdrawn from the earth

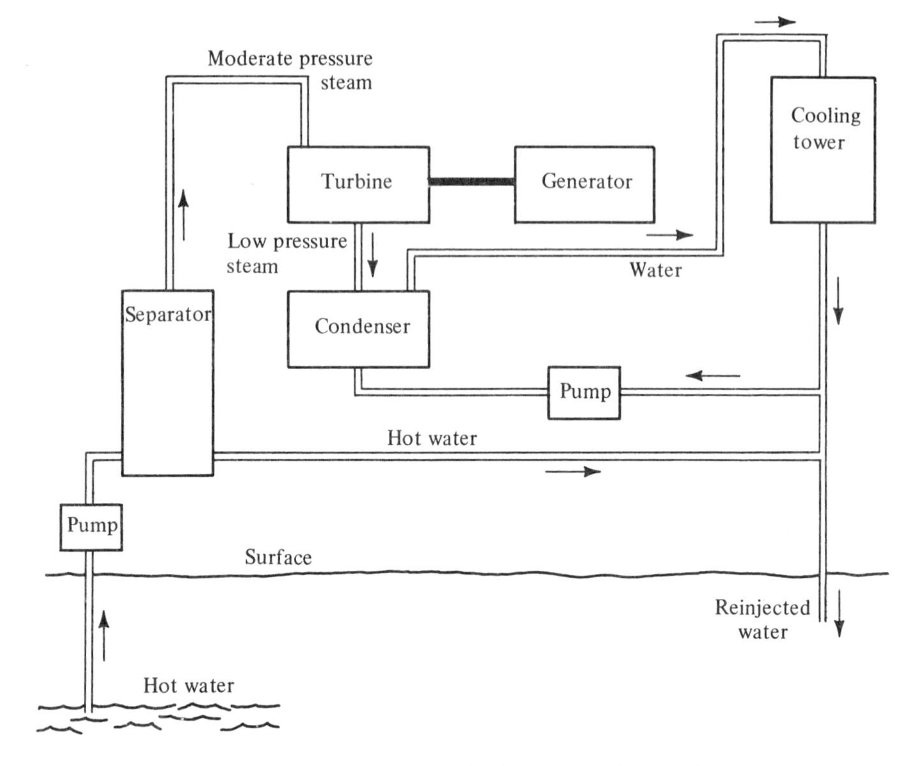

Figure 14-3. Steam power from flashing of geothermal hot water

is enormous. Consider a relatively small power plant, say 100,000 kW. Although the amount of water required to be withdrawn is dependent on its temperature, it could easily be some 4 to 5 tons per second. This is approximately 130 to 160 cubic feet per second. The continuing removal of such vast quantities of water from the earth may cause ground subsidence of appreciable magnitude. Since much of the geothermal water exists in unstable geological regions, there is some feeling that the withdrawal of very large quantities of water may be instrumental in causing earthquakes.

In general, geothermal water has a higher degree of salinity than does fresh water. For example, as mentioned earlier, some of this water may have a salinity six times that of ocean water. Obviously, the vast quantity of the spent water left over after power production cannot be dumped into streams but must be reinjected into the earth.

If this spent water can be reinjected in the right place, it will not only solve the problem of its disposal but will also minimize the problem of ground subsidence and possible earthquake production. The determination of the location of the correct point of water injection is not an easy one. If the water is injected too near the point of withdrawal, it will cool the water existing in the ground,

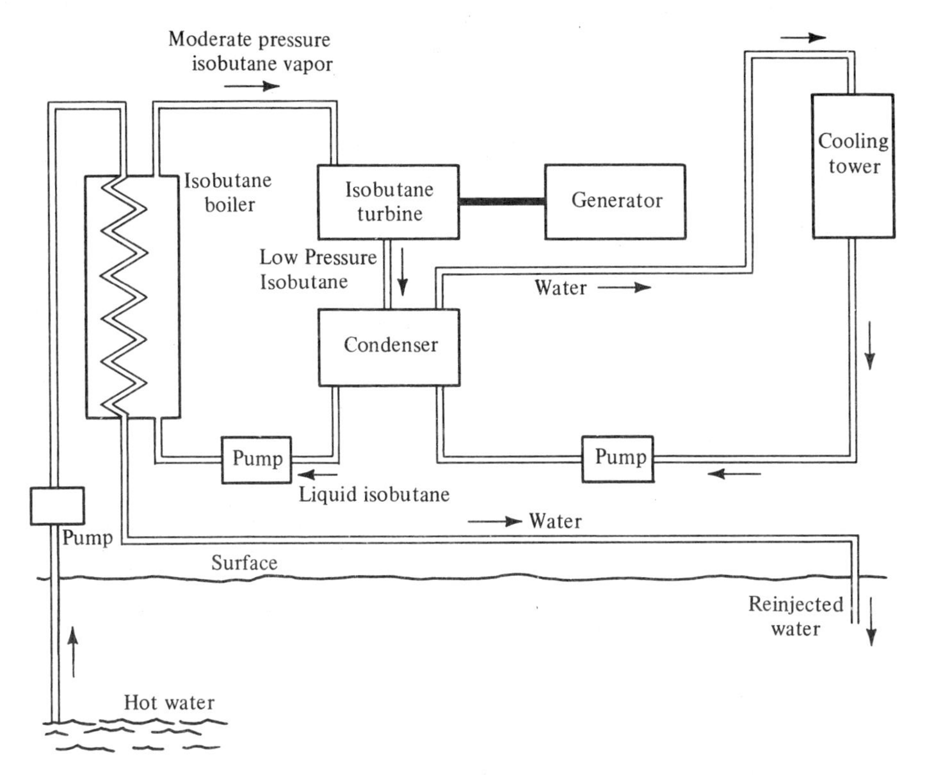

Figure 14-4. Power from geothermal hot water using isobutane

thus interfering with the delivery of hot water. If the water is injected too far away, it may not be of sufficient help in minimizing the subsidence problem.

There are, apparently, large amounts of geothermal hot water in the Imperial Valley in southeastern California. Much of this water has a high temperature, ranging up to 500°F. However, the salinity is very high, causing excessive scale formation in heat transfer devices. Since there is a critical water shortage in this area, it has been proposed that the energy remaining in the warm geothermal water after energy has been extracted for power production be used to desalinate some of the water. Because of the high salinity, it may also be feasible to recover some of the minerals in the highly concentratee brine left after the desalination process. There would be reinjection of the remaining brine. Consideration is being given to piping in ocean water 75 to 100 miles to make up for the water which has been desalinated. If this is done, the total water put into the ground will be equal to the amount originally removed, thus preventing subsidence.

The economic feasibility of geothermal hot water power plants has not been established since such plants have not been built. Those who have been working extensively in this area feel that such plants will become practical, particularly where the water temperature is high and the salinity is relatively low. This will be especially true if energy shortages increase and prices rise significantly. There is a wide range in the estimates being made as to the extent of the use of geothermal hot water in the next two or three decades. It should be recognized that estimates have been made on the basis of data incomplete for reasons which include

1. The total amount of geothermal water available for power production is not known and will not be known until very extensive geological surveys are made.
2. The actual costs of the total plant, including means of water reinjection, are not known and will not be known until at least one plant is made.
3. Much of the potential sources of geothermal hot water lie in federal lands. It is difficult to predict how readily this water will be made available for power production.

14.4 ENERGY FROM HOT ROCKS

Of all the energy stated earlier as possessed by the earth up to a depth of 6.5 miles, 99 percent exists in hot rocks, the rest in hot water and steam. Although hot water and steam are found in relatively few areas of the country, hot rocks can be found at depths in all regions. In many areas in the western part of the country temperatures as high as 450°F to 500°F are encountered at depths of 15,000 feet. In areas of non-volcanic origin temperatures ranging from 700°F to 800°F are encountered at depths of ten miles.

The common method of using the energy of hot rocks is to inject water which will be heated by them. Vast quantities of water must be heated to provide the energy required for even relatively small amounts of power production. The amount of water to be heated is dependent on several factors, such as temperature of the rocks, the temperature to which the water is to be heated, and the temperature at which the exhaust vapor can be condensed in the condenser. For illustration, assume that 100,000 kW of electrical energy is to be produced. Further assume that the water is to be heated by hot rocks to 400°F and assume that the heated water boils a secondary liquid, such as isobutane, at 300°F. If the condensing temperature in the isobutane condenser is 100°F it will take approximately 45 to 50 tons of water to be heated per second by the hot rocks.

To heat this quantity of water a very large heat transfer area of the rock will be required. Because of the number of variables involved it is difficult to determine even approximately the area of hot rocks required for this heating. (It should also be noted that although the temperature of the rock at 15,000 feet may be as high as 450°F, the rock in immediate contact with the water will be cooled to a significantly lower temperature and thus will not be able to provide as much heat per unit area of the rock). The author estimates that, for the given power output of 100,000 kW, at least 25 to 50 acres of rock surface area may be required.

Two methods have been proposed to obtain this very large surface area for the hot rocks. In one method, a hole is drilled down deep enough to reach rock at the desired temperature. Then, with an explosive, either conventional or nuclear, a cavern is created at the bottom of the hole. Water is introduced into the hole, heated by contact with the very large surface area of the cavern and then removed through a second hole drilled in a remote part of the cavern. (See Figure 14-5.) The heated water produces power in the manner described above for geothermal water. (See Figure 14-4.)

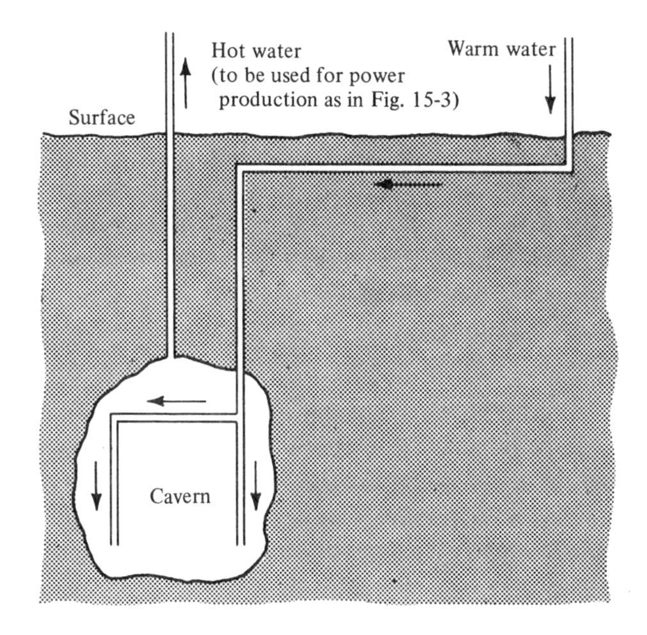

Figure 14-5. Use of caverns in hot rocks for power production

A second method of creating sufficient rock surface area also involves drilling a hole to the desired depth. Water, under high pressure, is introduced into the hole, causing the formation of one or more large radiating dish-shaped cracks in the rock. This method, known as hydraulic fracturing, has been used successfully in the oil industry to break up and release pockets of oil. A second hole is drilled into a remote area of a crack. Water is introduced into one hole, heated, and removed from the second hole to be used for power production as before. Thus far, consideration has been focused on using this method in impermeable rock only. Even in this type of rock, there is a possibility of sufficient underground leakages to prevent maintaining the water at the desired pressures. The introduction of cold water into the hot fractured rock will cause additional fracturing, created by the thermal stresses set up. This action is very desirable since it provides additional heat transfer area.

The Los Alamos Scientific Laboratory has been engaged since 1970 in studying the potential of hot rocks for power production. Several borings have been drilled near Los Alamos, New Mexico. In some cases, caverns have been created to provide the area required for heat transfer to the water. Studies have been made of such factors as rock temperatures, water loss through the rock, the increase in salinity of the water, and the decrease in rock temperature with time. In addition, with the cooperation with the U.S. Geological Survey, a survey is being made to determine the magnitude and distribution of economically accessible hot dry rock resources through the United States.

At present, for a short period of time, approximately 3 to 5 thousand kW of energy have been produced at one site. There are plans to develop a new downhole system in 475°F to 525°F rock which will be capable of developing 20 to 50 thousand kW thermal. The results obtained from this system by 1986 should indicate the economic feasibility of generating power from hot rocks.

It is possible that these studies will show that it is economically feasible to develop power from hot rocks in those parts of the country where the magma is relatively close to the surface. In other parts of the country, rock temperatures at reasonable depths may be sufficiently high to make it desirable to use their energy for heating buildings and for supplying heat at relatively low temperatures for industrial processes.

Whether the energy from hot rocks is to be used for power production or for heating, it should be recognized that unless very large areas are provided for heat transfer, the heat removed by the water from the surface of the rock will be greater than that which can be conducted through the rock to its surface. Thus, after a period of time—perhaps 30 to 50 years—the surface temperature of the rock will become too low to make it useful.

Although the economic feasibility of developing power from hot rocks has not been established, the very large potential that hot rock offers should be recognized. It has been stated* that a cubic mile of rock at 350°C (662°F) when cooled to 250°C (482°F) will yield enough energy to provide an electrical output of 3,700,000 kW for 30 years. Thus, 125 cubic miles of this hot rock could supply sufficient energy to satisfy the total present electrical demands in this country. A considerable amount of energy would be required initially to construct the energy removal system. Information is lacking as to how large this energy is when compared with the deliverable energy.

14.5 SUMMARY ON GEOTHERMAL ENERGY

The vast amount of geothermal energy stored in the earth's interior offers the potential for satisfying a major portion of our future energy needs. Three modes of using this energy are (1) geothermal steam, (2) geothermal hot water, and (3) energy in hot rock.

Only one area in this country is known to produce sufficient geothermal steam for large scale power production. This area, the Geysers region north of San Francisco, is now producing about 0.1 percent of the total electrical energy

* *Geothermal Energy Resources and Research*—Hearings before the Committee on Interior and Insular Affairs, U.S. Senate, June 15 and 22, 1972, p. 318—U.S. Government Printing Office.

generated in this country. Additional units are being added. It has been demonstrated that it is economically feasible to use this steam for power production.

It is recognized that there are many areas in the country, particularly in the western part of the country, where large quantities of geothermal water may be found. In some cases, the water temperature may be at least as high as 400°F to 500°F. It is believed that the total amount of geothermal water having the potential for power production is many times the presently known supply. However, extensive explorations must be made to ascertain the total potential of geothermal water.

Basic information is lacking concerning the extent of geothermal water, its temperature, and its salinity. Lacking also is information about the nature, temperature, and the amount of hot rock near the surface. Furthermore, the technology for using geothermal energy (except for steam) has not been fully developed. Hence there is a very wide range in the predictions of the amount of geothermal energy which will be produced in the coming decades. Some of those who have carefully studied this situation believe that no significant amount (less than one percent) of our electrical energy supply will come from geothermal sources. On the other hand, in 1972 in *Assessment of Geothermal Energy Resources,** it was estimated that at least 19,000 megawatts of geothermal generating capacity could be installed in the United States by the year 1985. This would represent over 2 percent of our total electric generating capacity. This same publication stated the possibility that by the year 2,000, geothermal energy could provide approximately 14 percent of the total electric generating capacity. Present indications, however, show that there is practically no possibility that geothermal electric power production will come close to the predicted 19,000 megawatts by 1985.

In *Geothermal Energy as a Source of Electricity,* a 1980 U.S. Department of Energy publication, the author, Ronald Di Pippo, gave a detailed discussion of the various activities in geothermal electric power production in this country and throughout the world.

In general, there may be serious environmental problems connected with the use of geothermal energy. Both geothermal steam and geothermal water may be corrosive. Both may contain poisonous gases, particularly hydrogen sulfide. The removal of some of these gases may be difficult. Geothermal water may have a high degree of salinity, in some cases as high as six times that of ocean water. Thus the water, after giving up energy, cannot be disposed of indiscriminately. The removal from the earth of vast amounts of geothermal water may cause serious subsidence problems and could conceivably trigger earthquakes in the geologically unstable regions where geothermal water is found in large quantities. Reinjection of the spent water could solve not only

---

* Committee on Energy Research and Development Goals, Federal Council for Science & Technology, *Assessment of Geothermal Energy Resources,* Washington D.C.; GPO, September 25, 1972.

the problem of its disposal but also help to minimize ground subsidence, provided that the proper place for reinjection can be determined.

Because of the great potential of geothermal hot water and of hot rock for meeting a large portion of our future electrical needs, it is essential that both governmental agencies and private industry investigate thoroughly the economic feasibility of using these two energy sources. If found feasible, much work will be required for geothermal energy to make a significant contribution towards meeting our energy needs. It is essential that extensive efforts be initiated immediately.

SUPPLEMENT A

ILLUSTRATIVE EXAMPLES

As stated in this chapter, enormous quantities of geothermal hot water are required to produce a given amount of electrical energy. This may be illustrated in the following example.

Example 14-1. It is desired to produce 100,000 kW of electrical energy. Geothermal water is available at 450°F. It may be cooled to 360°F, thus boiling a secondary fluid at a temperature of 350°F. The secondary fluid vapor passes through a turbine and is condensed at 100°F. The efficiency of power production by the secondary fluid is 65 percent of that of a Carnot cycle engine operating under the same conditions of temperature. Assume that the specific heat of the hot water is 1.1 Btu/lb-°F. Determine the amount of water required per minute to heat the secondary fluid.

Solution. Using Equation 1-2, the Carnot cycle efficiency is

$$\frac{350-100}{350+460}=31.8\%$$

The actual efficiency $\eta=0.65(31.8)=20\%$. Since

$$\eta=\frac{W}{Q_s},$$

the heat supplied per kWh is

$$\frac{3413}{0.2}=17{,}070 \text{ Btu}$$

where 3413 Btu is the equivalent of 1 kWh. The heat supplied per lb. of water, $Q=mc(T_2-T_1)$

$$1\times1.1\times(450-360)=99 \text{ Btu}$$

The pounds of water required per kWh equals

$$\frac{17{,}070}{99}=172.5 \text{ lb per hr or } 2.88 \text{ lb per min}$$

For 100,000 kW, the water required equals

$$2.88\times 100{,}000=288{,}000 \text{ lb/min or } 144 \text{ tons/min} \qquad \text{answer}$$

The vast heat transfer area of dry rock required to heat water for power production may be illustrated by the following example.

Example 14.2. Assume that the cooled water in Example 14.1 is to be injected deep in the ground to be heated by hot rocks. Also assume that the water temperature is 350°F before injection and that it is heated to 460°F by the hot rocks. Determine the heat transfer surface required.

Solution. The general equation relating heat transfer and area is

$$Q=hA(T_2-T_1)$$

Where Q is the heat transfer Btu/hr, h is the heat transfer coefficient, Btu/sq.ft.-hr-°F, and T_2-T_1 is the temperature difference, °F.

The heat picked up per lb. of water, using Equation 2-16, is $Q=mc(T_2-T_1)$

$$1\times 1.1(460-350)=121 \text{ Btu}$$

The total heat picked up equals the heat picked up per pound multiplied by the pounds per hour. Thus

$$Q=121\times 288{,}000\times 60=2{,}091{,}000{,}000 \text{ Btu/hr.}$$

The value of h varies greatly, particularly with the nature of the rock surface. Assume it to be 200 Btu/sq.ft.-hr-°F. Initially the rock temperature will be much higher than that of the water. However, the surface temperature will drop rapidly. Assume that the temperature difference stabilizes at 25°F. Then,

$$A=\frac{Q}{h(T_1-T_2)}=\frac{2{,}091{,}000{,}000}{200\times 25}=418{,}000 \text{ sq. ft. or } 9.6 \text{ acres} \qquad \text{answer}$$

SUPPLEMENT B

QUESTIONS AND PROBLEMS

14-1. Same conditions as in Example 14-1 except that the geothermal water is available at 400°F and that it may be cooled to 310°F, thus boiling the secondary fluid at 300°F. Calculate the tons of water required per minute.

14-2. Same conditions as in Example 14-1 except that the geothermal water is available at 300°F and that it may be cooled to 230°F, thus boiling the secondary fluid at 220°F. Calculate the tons of water required per minute.

14-3. Using the results of Example 14-1 and Problem 14-1 and 14-2, draw conclusions as to the possibilities for the use of geothermal water having a temperature much below 300°F to produce power.

14-4. Assume that water in Example 14-2 is to be heated by an isolated hot rock mass. Assume also that the rock will be cooled from 600°F to 480°F and that the rock has a specific heat of 0.2 Btu/lb.-°F. Determine how many tons of rock are required to heat up the water for a period of one day.

14-5. As stated in this chapter, there is a very wide range in the predicted amount of geothermal power which will be produced in this country in the next two or three decades. Give some reasons for this wide range in the predictions.

14-6. It presently requires 250 gallons of oil to heat a home in December. The oil has a heating value of 140,000 Btu per gallon and is burned with an efficiency of 70%. Geothermal hot water at a temperature of 140°F can be pumped up from a considerable depth in the ground and can be cooled to 100°F. Take the specific heat of the water to be unity. Determine the average gallons of water required per hour to replace the oil heat. (There are 7.48 gal./cu.ft.).

14-7. List factors to be considered before deciding to heat the home in Problem 14-6 by geothermal water.

15

Marine Energy

15.1 INTRODUCTION

As stated in Chapter 11, the oceans of the world receive vast amounts of energy directly from the sun. The oceans also receive an appreciable amount of energy from the winds and from gravitational forces. Some of the incident solar energy is reflected back into the atmosphere and other energy is dissipated by radiation from the surface of the oceans by virtue of their temperature. Much more energy is used in vaporizing water which is ultimately returned to the earth in the form of rain and snow.

Because of temperature gradients in various parts of the ocean, together with wind action and forces created by the earth's rotation, currents are created, some warm and some cool. The warm currents tend to keep the temperature of adjacent land areas above that normally associated with their latitudes. As an example, the Gulf Stream maintains the temperature of western Europe much higher than normal for land that far north. On the other hand, the cool ocean currents maintain the temperature of adjacent land areas below that normally associated with the given latitude.

Of all the various forms of energy possessed by the oceans, five seem attractive for power production:

a. Tides.
b. Wind waves.
c. Ocean currents.
d. Thermal gradients.
e. Salinity gradients.

Each of these will be examined as to their potential as well as to the problems associated with their use.

15.2 TIDAL POWER

The gravitational force of the sun and the moon tend to cause the waters of the oceans to lag behind as the earth rotates, giving rise to what is known as a tidal bulge. This results initially in the low tides and then in the high tides. After a low tide, the oceans surge in toward the shore. For much of the earth's shoreline, the difference between high and low tide may be five to six feet. However, in certain regions, bays are located so that the tides funnel into them, increasing the height of water at high tide. In a few of these bays throughout the world, the difference between high and low tides may be as much as 50 feet.

The idea of using the difference between high and low tide for power generation is not new. Over a hundred years ago, a small spice mill near Boston, Massachusetts developed a small amount of power, intermittently, by installing a water wheel in a sluiceway connecting the ocean with a storage basin. In the early 1930's, very serious consideration was given to the construction of a very large tidal power plant in the Passamaquoddy Bay, an arm of the Bay of Fundy, located between Maine and Canada. The difference between high and low tide is around 35 to 40 feet. Although detailed plans had been developed for the project, it was decided that it was not economically feasible to build. Since that time, the idea of a tidal power plant at this site has been reconsidered periodically. However even today, with high power costs, it still is not economically sound to develop a tidal power plant at this site.

At present, only two significant tidal power plants are in operation. One of these, a 400 kW unit, is located in the U.S.S.R. The second one, with a capacity of 240,000 kW, is located in France on the Rance River, near St. Milo. Thus, it has been demonstrated that it is technically feasible to derive power from the tides, but the economic feasibility of doing so has not been demonstrated.

It is recognized that the tides possess vast amounts of energy. We have the technology for using this energy. Why is it that we are not doing so? It was pointed out in Equation 10-3 that large quantities of water are required together with a relatively high head to produce sizeable amounts of hydropower. The same principle applies to tidal power. Only in a very few regions is the difference between high and low tide as much as 15 feet, with an average working head only about 7.5 feet. For these conditions, extremely large amounts of water are required for even a moderate power output. In addition, a very large storage basin is required. The top of this basin should be at high tide elevation. Under these conditions, it does not appear that it will ever be economically sound to produce tidal power in the United States except at the Passamaquoddy Bay and possibly at the Turnagain Bay in Cook Inlet, Alaska.

15.3 WIND WAVES

It is apparent by watching ocean waves dash onto the shore that they possess much energy. This is particularly true in times of violent storms.

Various proposals have been made to harness this energy, some with devices being located close to the shore and others out in the open ocean. Some of these devices were actually constructed as early as the first part of this century, demonstrating their technical feasibility on a small scale basis. For economic reasons, no large scale wave machines have been built as yet. However, extensive work is going on, particularly at the University of Edinburgh, to harness the energy of ocean waves near the Northern coast of the British Isles. In this region winds blowing across the Atlantic produce wave energies averaging about 80 kW per meter of wave front, with the energy being over 320 kW per meter for a considerable portion of the time. Preliminary model basin studies indicated the feasibility of using this energy. A 50 kW unit is now being constructed. Unlike many power producing devices which require large units to reproduce a significant amount of power, large amounts of wave power can be obtained by simply using many small units. Thus, if favorable operating data is obtained from the first unit, additional units can be mass produced, greatly lowering the initial costs. It is too early to predict the possibilities of using wind energy off the Northern coast of the British Isles.

The prospects of using ocean wave energy in this country are far from being as good as those in Great Britain. Even in the most favorable region, the Washington and Oregon coasts, the average wave energy ranges from 40 kW per meter of wave front in the winter down to less than 20 kW per meter in the summer. Along the Atlantic coast, the average wave energy averages between 3 and 5 kW per meter of wave front. Because these energies are so small, it does not appear that it will be economically feasible to harness them.

15.4 OCEAN CURRENTS

It is recognized that the various ocean currents such as the Gulf Stream, the Labrador Current, and the Humbolt Current possess large amounts of energy by virtue of their velocity. For example, the Gulf Stream at one point off the coast of Florida has a central core width of about 16 miles and a depth of 750 feet. Its mean velocity varies from 3.34 miles per hour in winter to 5.18 miles per hour in summer. Thus, at the mean velocity, the kinetic energy of this current is over 20,000,000 kW.

One group has been working for several years to use the kinetic energy of the Gulf Stream. It is stated that they have spent so far over a million dollars for research and design of an underwater turbine set-up to produce power from the Gulf Stream current. It appears that a turbine unit having a diameter of 245 feet and rated at 5,000 kilowatts would be the optimum size. They have built models which were pulled through water and also tested in the David Taylor Model Basin. They estimate that power can be produced at costs comparable with those of many power companies. Of course these costs cannot be verified until units are actually built and tested. It should be noted that through this method of using the energy of the current the production of 200,000 kilowatts of electrical energy would decrease the energy of the Gulf Stream by less than one percent.

15.5 THERMAL GRADIENTS

In many parts of the oceans the difference in the water temperature at the surface and that at depths of a half mile or more may be 26° to 35°F. The Carnot principle indicates that when a temperature difference exists, there is a potential for power production. (See Chapter 1.)

The first known successful attempt to use this power potential was made in 1929 by a French engineer named Claude. He installed a 22 kW plant on the coast of Cuba. The plant worked but operating difficulties were great. In particular, it was difficult to maintain the very long pipe which brought the cold water from the depths offshore to the plant which was located on shore. Wave action, particularly during storms, broke the pipe.

In the simple system, the warm surface water is drawn into an evaporator. A vacuum pump is used to reduce the pressure in the evaporator to a very low value, thus permitting some of the water to boil. For example, when water having a temperature of 77°F is introduced into an evaporator where the pressure is maintained at 0.4158 psia, approximately 0.3 percent of it will flash into steam. The temperature of this steam will be 74°F. The steam thus formed flows through a turbine and hence into a condenser. The circulating water for the condenser is brought up from depths of a third of a mile or more. This cold water will condense the steam at a low temperature, maintining the steam pressure in the condenser and at the turbine exhaust at a sufficiently low value to produce work. For example, steam at 41°F will condense when the pressure is 0.12648 psia.

The simple thermal gradient system described above is not very satisfactory for large scale plants. This is primarily true because of the tremendous volumes occupied by the steam at such low pressures. At 74°F, the volume of a pound of steam is 763.5 cubic feet. At turbine exhaust conditions its volume may be close to 2,000 cubic feet per pound. In addition, the turbine work done by a pound of steam under these conditions is less than one tenth of that done in a conventional steam power plant. This means that at least ten times as much steam must be used for a given power output. This combination of a tremendous amount of steam and its very large specific volume means that the total volume to be handled for large output plants is prohibitive.

The alternative is to use the warm surface water to boil a secondary fluid such as isobutane or ammonia. The secondary vapor thus produced flows through a special design turbine and into the condenser where it is condensed by the cold water brought up from the ocean depths. (See Figure 15-1.)

To avoid the difficulties encountered with the very long cold water pipe when the plant is located on shore, it is proposed to situate the entire unit out in the oceans. There are two possible configurations of such plants that are being considered.

The more common concept is to submerge the entire system at a sufficient depth so that none of it will be disturbed by storms. When this is done, working quarters must be provided under water for the operating personnel. The

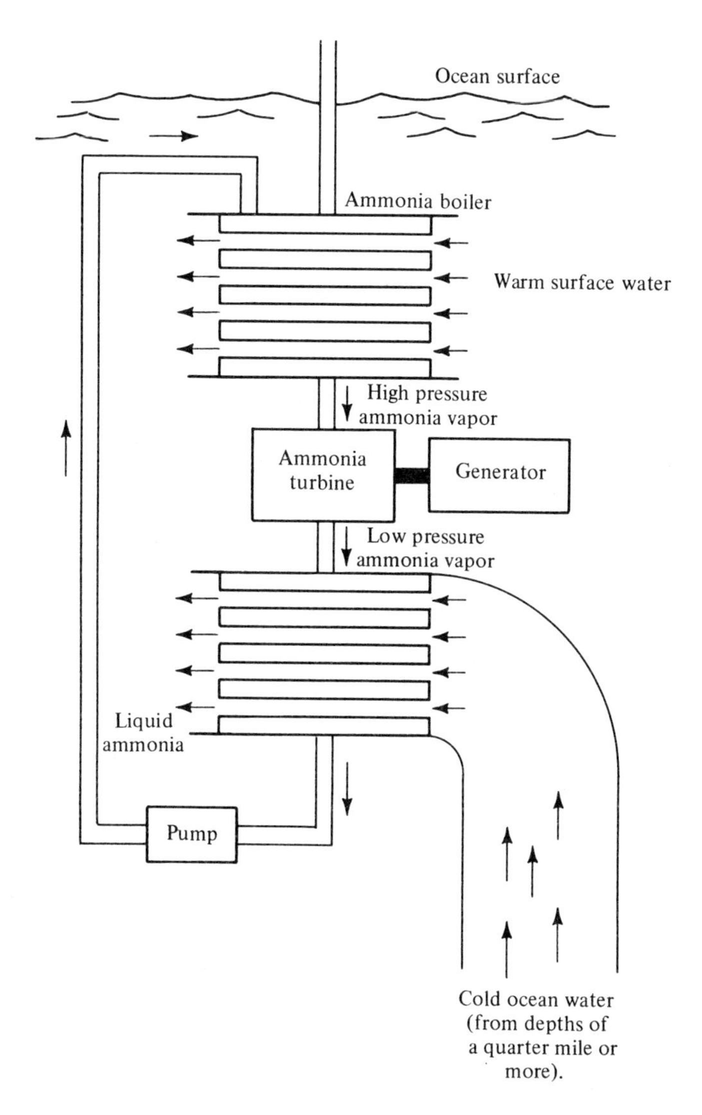

Figure 15-1. Elements of an ocean thermal gradient power plant

second approach is to build the system into a specially designed ship, with both the evaporator and condenser being placed beneath the bottom of the ship. The use of a ship for this purpose has to be restricted to non-hurricane prone areas.

A logical location for a submerged Ocean Thermal Energy Conversion (OTEC) plant in this country is in the Gulf Stream, particularly off the coast of Florida. Siting of such a plant north of Florida appears less attractive. As the stream moves northward, there is some decrease in the temperature of the water at the surface. A decrease in the temperature difference of only a very few degrees makes a significant difference in the efficiency and hence in the power output. Furthermore, the stream has some tendency to spread out and its surface temperature becomes somewhat influenced on the land side by the rivers of the region. Since there is a variation in river flow, there may be appreciable variations in surface temperature in parts of the Gulf Stream.

After a study of sea conditions in various parts of the oceans, engineers at the Johns Hopkins Applied Physics Laboratory selected a site some distance off the coast of Brazil as an excellent location for a ship type OTEC power plant. The difference in temperature of the surface water and that a depth of 2,500 feet averages 43°F. Furthermore, this area is not hurricane prone.

Several groups, with support from the federal government, have recognized the great potential of the warm surface water of the oceans for power production and are making extensive studies of the feasibility of using this energy. Several problems have been identified in this development. Since the working temperature difference is very small, the thermal efficiency is very low, perhaps 2.5 to 3.5 percent. This means that vast quantities of surface water are required to supply the heat to the secondary fluid and large quantities of cold water must be brought up from the ocean depths to be used in the condenser. Thus, even for a relatively small unit, say 100,000 kW, the cold water pipe must be over 100 feet in diameter. Likewise, the heat exchangers must be very large.

A second problem concerns materials. Much information is available about materials suitable for ocean-going power plants. In general the materials needed to resist salt water corrosion are expensive. There is some hope that cheaper materials, such as an aluminum alloy, will prove satisfactory for the condenser since the cold water used in the condenser is less corrosive than the warmer surface water. Information is lacking, however, to permit making an accurate estimate of the useful life of such materials for this type of service.

A third problem connected with OTEC plants is that of using the energy produced. Some suggestions for doing this are

1. Transmit the electrical energy to shore.
2. Produce hydrogen by electrolysis of water.
3. Produce ammonia.
4. Produce aluminum from bauxite ore.

When it is feasible to locate an OTEC plant close to land, the electrical energy produced may be transmitted by underwater cables. Most OTEC plants, however, probably will be located too far offshore to use this method, unless the technology for underwater electric cables becomes more advanced than at present.

Much consideration is being given to the use of the electrical energy developed in OTEC plants to produce hydrogen by electrolysis of the ocean water. The hydrogen could be piped underwater to land or it could be liquefied and shipped to land in special refrigerated ships. There is a considerable range in the efficiency of producing hydrogen by electrolysis, but an average value seems to be around 50 percent. Even when the hydrogen is used for heating purposes, the heating effect may not be more than one-third of that of the electrical energy from which it was produced. When the hydrogen is used for electric power production, the power produced will probably be less than one-fifth of the electrical output of the OTEC plant.

It has been suggested that the electrical energy produced in a ship type OTEC plant be used aboard the ship for the manufacture of ammonia. The nitrogen required can be obtained from the atmosphere and the hydrogen from the water. Ammonia, a vital ingredient of fertilizers, is now made largely with natural gas. With the growing demand for fertilizers and the growing shortages of natural gas, it would appear that the ammonia produced by an OTEC ship would find a ready market.

Enormous amounts of electrical energy are required in the production of aluminum from bauxite. Much of our bauxite is imported, particularly from Jamaica. Thus, it would be easy to ship the bauxite to an OTEC ship and use the electrical energy developed there to produce aluminum. The aluminum thus produced could readily be shipped ashore.

In July 1980, the *S.S. Ocean Energy Converter,* a converted Navy oil tanker, was dedicated in Honolulu harbor. It represents a 41 million dollar investment by the Department of Energy. The primary purpose of this converter is to obtain performance data on the components of a complete unit. Although a relatively small unit, it will circulate 150,000 gallons per minute of surface water through 44 miles of tantalum tubing in the boiler to produce ammonia vapor for the operation of the turbine. A cluster of three 48-inch diameter polyethylene pipes descend 2,100 feet from the hull to bring up cold water for the condenser. Although this unit is relatively small, it should supply such information as to the difficulty in preventing scale build up on heat transfer surfaces.

Detailed calculations have been made of the anticipated costs of power production in OTEC plants. Many of these estimates show that OTEC plants are economically sound. Although a study of the *S.S. Ocean Energy Converter* will be of some help in estimating the power production costs of a full scale unit, it will be necessary to build somewhat larger units before a final conclusion can be reached as to the economic feasibility of generating large amounts of power in OTEC plants. Because of the growing energy shortages and rising energy costs, it is very desirable to continue investigations of the economic feasibility of a large size OTEC plant.

15.6 SALINITY GRADIENTS

When a semi-permeable membrane* is placed between waters of different salinities, a pressure difference, known as the osmotic pressure, is created

* A semi-permeable membrane will permit certain types of molecules to pass through it but will prevent the passage of other molecules.

across the membrane which causes fresh water to pass from the low salinity body of water to the higher salinity body. The flow of fresh water can be reversed by exerting a pressure on the high salinity water which exceeds the osmotic pressure. In fact, this is one method of desalinating salt water.

At the mouths of rivers discharging into the oceans, conditions exist to create the osmotic pressure by separating the fresh and salt waters with the use of a semi-permeable membrane. The magnitude of the osmotic pressure created is dependent on the salinity of the two bodies of water separated by the semi-permeable membrane and on their temperature. For water at 77°F, if a semi-permeable membrane were to be stretched across the mouth of a river to separate it from the average sea water (salinity 35 parts per million), the osmotic pressure would be about 23 atmospheres, which is equivalent to a head of approximately 785 feet of fresh water.

Several techniques have been proposed to use the osmotic pressure to develop power. One of these methods is illustrated in Figure 15-2. Two dams are erected, one at the mouth of the river and the second one a short distance out in the ocean. A pipe conducts the fresh water to a turbine where power is produced. The water leaving the turbine is discharged into the space between the two dams, (known as the buffer zone) from which it flows through a second pipe into the ocean. A semi-permeable membrane is placed at the end of

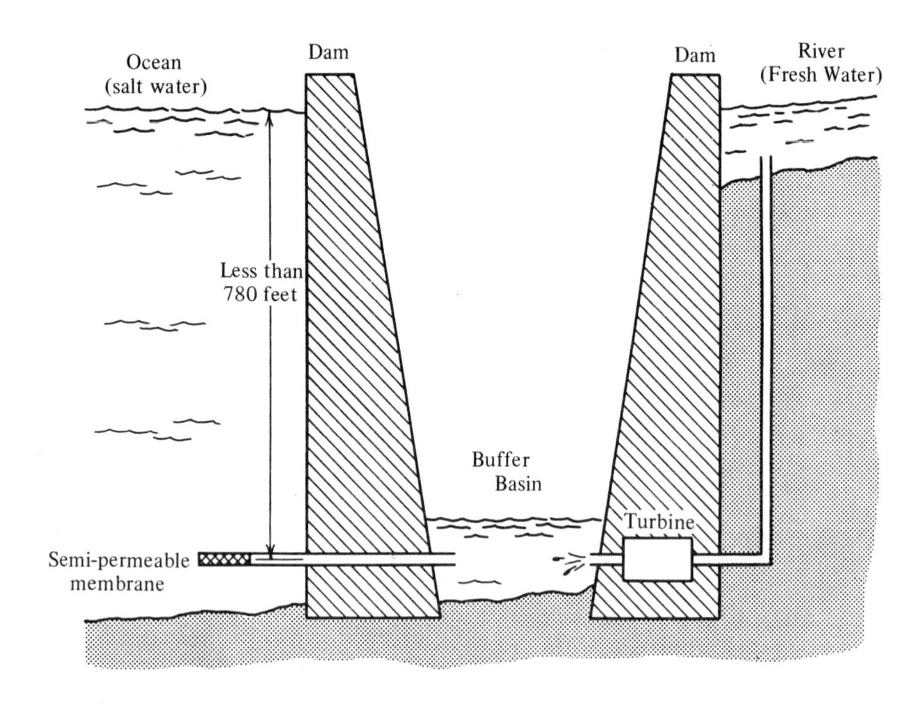

Figure 15-2. Power from salinity gradients

this pipe. Osmotic pressure will cause the fresh water to flow from the buffer zone into the ocean, provided that the water level in the buffer zone is sufficiently less than 785 feet below sea level so that there is ample head to overcome the resistance in the membrane and in the discharge pipe.

Because of the large amount of water discharged into the oceans by the aggregate of all the rivers, the potential of power production using osmotic pressure is high. The cost of using this energy source will vary, depending on the contour of the land at the mouth of the river. The damming of a river at its mouth could create several problems. Silt would collect behind the dam. Shipping would be interfered with. Movement of marine life between the river and the ocean would be stopped. Since no installations have been made using osmotic pressure, no information is available for accurate prediction of either the economic or technical feasibility of such a plant. Neither is it possible to estimate the adverse effects of this type of a plant. However, because of the high potential of producing significant amounts of power utilizing osmotic pressure, it seems desirable to carry out detailed investigations, perhaps by building a small plant.

15.7 SUMMARY ON ENERGY OF THE OCEANS

The oceans of the world receive vast amounts of energy from the sun. Much of this energy is available to help meet our energy needs. Unfortunately, the energy of the oceans is so diffused that it is difficult to use much of it. This is particularly true for ocean currents, although it is possible that it may become economically feasible to use the energy of certain ocean currents which have relatively high velocities.

Although it may be economically desirable to use the appreciable wave energy available off the northern coast of the British Isles, the wave energy adjacent to both our Pacific and Atlantic coasts does not appear to be sufficiently large to justify its use. With the possible exception of one or two locations, the difference in height between our high and low tides is not large enough to justify our development of tidal power.

Perhaps the type of ocean energy which shows the greatest promise for use is the thermal gradient energy. However, since no operating information is available for such systems, even small ones, it is very difficult to predict how extensively this energy will be used in the future. In any case, it does not appear that we will use a significant amount of ocean thermal gradient energy in the next decade or two.

Because of the osmotic pressure set up when fresh and salt waters are separated by a semi-permeable membrane, the potential for power production exists at the mouth of the rivers as they discharge into the oceans. However, since these types of plants have not been built outside of the laboratory, information is not available to predict their economic feasibility as well as the possible adverse ecological effects.

SUPPLEMENT A

ILLUSTRATIVE EXAMPLES

Example 15-1. It is proposed to develop tidal power in a region where it is possible to construct a storage basin 1 mile wide, 4 miles long and 3 feet deep. Assume that this basin is completely filled at high tide. As the tide recedes, water flows from the basin through a turbine. Assume that the entire contents of the basin flow out in a six hour period. Further assume that the average working height is 3 feet and the turbine efficiency is 88 percent. Take the density of sea water as 64 lb. per cu. ft. Determine the average horsepower produced.

Solution. The volume of the basin is

$$(4\times 5{,}280)(1\times 5{,}280)3=3.346\times 10^{8}\text{ cu. ft.}$$

The average mass rate of the outflow per minute is

$$\frac{3.346\times 10^{8}}{6\times 60}\times 64=59.5\times 10^{6}\text{ lb. per min.}$$

The horsepower equals

$$\frac{59.5\times 10^{6}}{33{,}000}\times 3\times 0.88=4{,}760\text{ hp} \qquad \text{answer}$$

Example 15-2. A 100,000 kW OTEC plant is to be installed in the ocean where the surface water temperature is 79°F. At a depth of 2,500 feet, the temperature is 42°F. The efficiency of the plant is 46 percent of that of the Carnot cycle plant for the given temperatures. Calculate the heat to be given up per second by the warm surface water.

Solution. The Carnot cycle efficiency is

$$\frac{T_h-T_c}{T_h}=\frac{79-42}{79+460}=6.86\%$$

The actual efficiency is

$$6.86\times 0.46=3.15\%$$

The heat supplied per second is

$$\frac{W}{\eta_t}=\frac{100{,}000\times 3{,}413}{0.0315\times 3{,}600}=3{,}010{,}000\text{ Btu per sec.} \qquad \text{answer}$$

Example 15-3. The surface water in Example 15-2 is used to boil a secondary fluid by cooling down 3°F. Its density is 64 lb. per cu. ft. and its specific heat is 0.95 Btu per lb. per °F. Calculate the cubic feet of water required per second.

Solution. The heat given up per pound of water is

$$3 \times 0.95 = 2.85 \text{ Btu}$$

The volume rate of water flow is

$$\frac{3{,}010{,}000}{2.85 \times 64} = 16{,}500 \text{ cfs} \qquad \text{answer}$$

SUPPLEMENT B

PROBLEMS

15-1. Same as Example 15-1 except the storage basin is 10 miles long, 1.5 miles wide, 5 feet deep and the average working height is 5 feet. Compare results and comment on the feasibility of tidal power generation for these conditions.

15-2. An ocean current is 50 miles wide, 10 feet deep and has an average velocity of 2 miles per hour. Determine the horsepower which can be developed from this current, assuming that 50 percent of the kinetic energy can be utilized. Comment on your results.

15-3. Find the amount of circulating water required for the condensers in Example 15-2 in cfs. Assume that all of the heat supplied which is not delivered as work is rejected in the condenser. The cold water increases 3°F in temperature in the condenser.

15-4. The velocity of the water in the pipe entering the condenser in Problem 15-3 is six feet per second. Determine the pipe diameter.

15-5. Same as Problem 15-3 except the surface water temperature is 76°F.

15-6. Discuss the desirability of funding for research and development for the various types of energies of the oceans. Should they all be funded at the same level?

15-7. In the discussion of using osmotic pressure for power production, it is suggested that dams be built at the mouth of rivers. Why is it not possible to obtain power directly from damming up the river rather than to use membranes?

15-8. Will any significant amount of energy be produced directly from the oceans in this century? If so, how?

15-9. The diameter of the pipe in Problem 15-4 can be reduced greatly if the water velocity is increased to, say, 18 feet per second. Is this recommended?

16

Shale Oil and Tar Sands Oil

16.1 OIL SHALE

Oil shale is described as a fine grained sedimentary rock containing an organic material known as kerogen. In the true sense, oil shale is not a shale. Generally, it looks like ordinary blackish or brownish rock. Frequently, however, it is structured in layers which may be colored yellow, red, or green by the waxy organic materials which are present. Although the kerogen is not oil, it does produce a form of oil when heated. There is no satisfactory explanation as to the origin of oil shale, although it is felt the kerogen was formed from vegetation.

Oil shale may be found in many countries of the world. In some countries it has been mined to a limited extent and oil extracted from it. There are very large quantities of oil shale in Brazil and still larger quantities in the United States. In all other countries the known oil shale resources are not sufficiently large to add significantly to the known supply of crude oil.

Particularly in mountainous or hilly regions, the oil shale may occur as more or less barren cliffs. In some cases it may be found as outcroppings. In other cases the oil shale is covered by an overburden of various thicknesses. Some of the richest oil shale lies at a considerable depth below the surface. Frequently, the oil shale is found in veins, some being several hundred feet in thickness. Veins as thick as 2,000 feet have been reported. Particularly in the thicker veins, there may be considerable variation in the concentration of the kerogen throughout the vein.

The richness of oil shale is reported in terms of the oil potential, in gallons per ton of shale. Most of the oil shale has a potential of less than 50 gallons per ton. Vast quantities of shale contain less than 10 gallons per ton. Although the total potential of this oil-poor shale is tremendous, the oil potential per ton of shale is so low that it is not normally considered as a recoverable resource. As with other minerals, it is very difficult to estimate the total potential of oil shale. By far the largest amount of oil shale is found in the Green River basin in

Colorado, Utah and Wyoming. An extensive discussion (517 pages) of this oil shale, entitled *An Assessment of Oil Shale Technologies*, was prepared by the Office of Technology Assessment of the United States Congress and printed in 1980. This publication presents an estimate of the total amount of oil present in the shale, reviews the development activities to date, and discusses the problems associated with shale oil production. It discusses in considerable detail the various effects on the region by the production of various amounts of oil ranging from 200,000 to 1,00,000 barrels per day.

This publication shows the potential shale oil in seams at least 100 feet thick and yielding at least 30 gallons per ton to be 418 billion of barrels. In addition, it lists a potential of 1,400 billion barrels of shale oil to be found in seams at least 15 feet thick and yielding at least 15 gallons per ton.

It is difficult to estimate how much of this shale oil can be recovered. It appears that as soon as shale production methods are perfected it will become economical to produce shale oil from the very rich oil shale. Then, as the price of petroleum increases, it should become economically feasible to develop less rich shale oil deposits. It seems possible that at least 100 billion barrels of shale oil can be produced. If oil prices continue to increase rapidly one estimate places the amount of economically recoverable shale oil at a trillion barrels. (It should be noted that at present we are using over 6 billion barrels of oil annually). However, many serious problems involved in shale oil production must be solved before shale oil can be produced in large quantities.

16.2 SHALE OIL PRODUCTION

A few small shale oil industries were developed in the United States in the 1850's in the Appalachian Region and in the Ohio Valley. Oil and illuminating gas were produced. However, these enterprises were discontinued when Drake drilled his first oil well in 1859, producing oil at a much lower price than shale oil.

Because of an oil shortage during World War I, efforts were initiated to develop the shale oil industry. However, the discovery of vast oil fields, particularly in Texas, cut short further development of the shale oil industry at that time. In 1944, Congress passed the Synthetic Liquids Fuel Act, directing the Bureau of Mines to conduct research and development work on methods of producing commercial oil products from oil shale and coal. Even as early as 1945 over 2,000 patents had been issued in this field. Unfortunately, very few of these ideas got beyond the drawing board and fewer still were ever tried out in working models. In the last 35 years several oil companies have constructed pilot plants at various locations. These range in size up to 1,200 tons of shale per day. Occidental Petroleum, Atlantic Richfield, and Union oil have spent millions of dollars experimenting with shale oil extraction. The Chairman of Occidental Petroleum believes that his company will be able to begin commercial production by 1985 at a cost competitive with domestic crude oil. In May of 1980, Exxon paid Atlantic Richfield 400 million dollars for its share of the

Colony oil shale project in Colorado and now plans to spend 500 billion dollars over the next 30 years to build 150 installations on Colorado's Western Slope. If this is done, it is estimated that the output by the year 2010 will be as high as 8 million barrels of oil per day. Since this amount is approximately 45 percent of the amount which we are now using it may appear that the source of our future oil needs will not be a major problem. However, it should be realized that many predictions have fallen way short of their anticipated goals.

At first, efforts to obtain oil from the shale involved mining it and then retorting it. The first step in the above ground production process is to mine the oil shale, generally from an underground mine. The methods developed for mining coal can be applied here. There are two primary differences. Normally the oil shale veins are very much thicker and hence the problem of roof support is more difficult. Oil shale is harder than coal and hence is more difficult to mine. After mining, the oil shale is crushed and then fed into a retort. There is a considerable difference in the details of the various retorts. The crushed shale may be admitted at the base of the retort and forced upward. See Figure 16-1. The shale may also be fed in at the top and allowed to descend towards the bottom. In either case, combustion with a deficiency of air raises the temperature to approximately 900°F. At this temperature most of the kerogen is vaporized. Condensation of some of the vapors may occur in the cooler

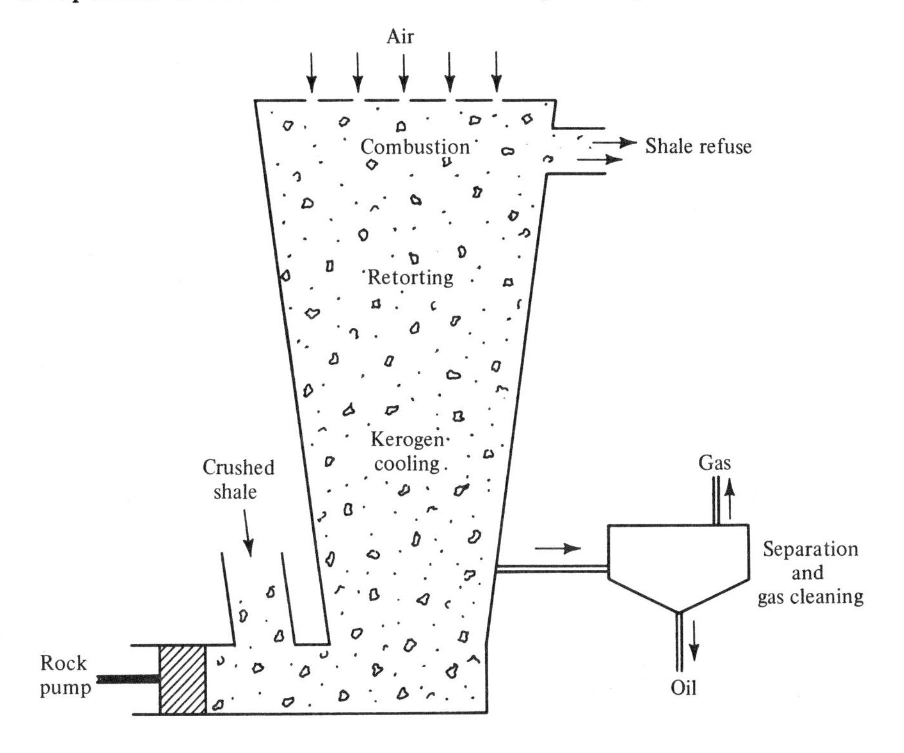

Figure 16-1. Oil shale retort

parts of retort. The vapor-liquid droplet mixture is removed from the retort and further cooled to produce more liquid. The vapors not condensed may be fed back into the retort to be burned, thus providing much of the necessary heat. The rest of the heat is obtained by burning the kerogen which does not vaporize. The kerogen liquid can be refined to produce a form of oil. Before refining, it is a black viscous liquid which, when cooled, will solidify about 90°F into a jelly-like substance. This makes it, of course, very difficult to handle.

Recently a considerable amount of effort is being devoted to in-situ (underground) production. In one such operation, shafts are dug down into the oil shale bed. Horizon tunnels lead from the shaft. At various points in these tunnels caverns are created, which are topped by retorts, reported to be as high as 30 stories in some cases. Blasts are used to break up the adjoining shale and to fill up the cavern. Oil is introduced at the top of the cavern and ignited. The resulting high temperature drives out the kerogen which flows down to the bottom of the cavern where it can be recovered. Still left in the shale is a carbon residue. Air and steam are introduced to burn this carbon residue in a controlled manner to ultimately heat up all of the shale rubble in the cavern, thus driving off the kerogen.

Although many problems are associated with the in-situ method, it does not require large quantities of water and the residue shale disposal is not a problem. Recently, another approach has been made to in-situ production of shale oil which uses radio waves or microwaves to heat the shale sufficiently to release the kerogen. It is too early to predict the possible commercial success of this method.

It has been demonstrated by the use of pilot plants that we have the technology to produce oil from oil shale. About 80 percent of the oil shale, particularly the oil rich shale, lies on government owned lands. Hence, the policy of the federal government will determine, to a very large extent, the future development of shale oil. To encourage this development, in 1974 the government requested bids for leases on six tracts of oil shale lands, each over 5,000 acres in size. Bids were received from consortia of oil companies for the two tracts in Colorado and the two in Utah, but none for the two in Wyoming. The highest bid was for over 200,000,000 dollars for one of the tracts in Colorado. For some time there has been a legal dispute between the State of Utah and the federal government as to ownership of the land claimed by the federal government. This dispute has prevented the development of the two shale oil tracts in Utah. Very little appears to have been done on the two tracts in Colorado. A few years ago, the price of oil was not sufficiently high to justify development of shale oil at that time. With today's oil prices and the prospect of still much higher prices, it now seems that shale oil can be produced at competitive prices. However, billions of dollars will be required to construct and operate full-size shale oil plants. Although the oil companies are receiving high profits and have vast resources they probably will be unable on their own to finance full scale plants. Because the economics of shale oil production have not been established it is difficult to obtain from investors the vast sum of money required.

It is difficult to predict the magnitude of shale oil production in the future. Several factors will govern the future of shale oil. Some of these are

1. Government policies
2. Ability to raise the vast sums required
3. Future price of oil
4. Ability to solve the many problems associated with shale oil production.

If present activities are expanded at the expected rate, shale oil production could reach as high as 300,000 barrels per day by 1990 and perhaps at least 1,000,000 barrels per day by the year 2,000. For a basis of comparison, we used about 17,500,000 barrels per day in 1979.

16.3 PROBLEMS ASSOCIATED WITH SHALE OIL PRODUCTION

There are many difficult problems associated with shale oil production. The major ones will now be discussed.

1. The financial problem has been discussed already to some extent. Questions must be raised as to whether the oil companies are to finance shale oil production. If so, how are they to obtain the sums of money required? If not the oil companies, is there any hope of any other financial group being able to obtain the vast sums of money required? Or has the federal government demonstrated that it can make such a development efficiently and economically? In any case, are we, the consumers, willing to either pay the price for the shale oil or the taxes to develop it?

2. We have the technical knowledge necessary to mine the shale, to crush it and to retort it. It will require, however, much engineering know-how to adapt this knowledge to full-scale operation in making the mining of the shale safe for the miners and in making the crushing operation and the retorting operation non-destructive of the environment. If the shale is to be obtained by strip-mining, care must be taken to restore the land to satisfactory conditions.

3. It will be necessary to dispose of huge quantities of spent shale. The spent shale has a volume approximately 12 percent greater than the original volume. The spent shale could be placed in the holes created when strip-mining is employed. Much of the countryside where the oil shale is located is desolate and has very few inhabitants. It is proposed to dump the spent shale into the valleys of the region. When mixed with sufficient water, the spent shale has a tendency to set up like a weak cement, thus becoming stabilized. However, rains will have a tendency to leach the spent shale and carry harmful chemicals down into the streams of the region. Since the rainfall in this region of the country is low, it may be necessary to import water for revegetation.

4. The oil produced by retorting of the shale could be refined immediately to make various satisfactory oil products. However, refining of the liquid re-

quires enormous quantities of water which generally is not available in the oil shale region. One alternative is to crack the liquid sufficiently to so reduce its viscosity so it can be pumped several hundred miles to sites where sufficient water is available for refinery operation.

A second alternative is to use the energy of the oil produced from oil shale for mineral recovery from the spent shale. Among other minerals which oil shale possesses, it contains nahcolite (from which soda ash is obtained) and dawsonite (from which alumina is obtained). It has been estimated that one square mile of certain very thick vein oil shales contains five times the 1972 comsumption of primary aluminum in this country. As the price of imported aluminum ore (bauxite) increases, it may become desirable to mine some oil shale for its aluminum and perhaps for its soda ash potential and to use the contained shale oil to produce the required energy.

5. A very critical problem associated with shale oil production is shortage of water. Water is required in the mining operation, in crushing and in retorting. Water is also required for refining, even if the oil is simply cracked to permit its flow in pipelines. Much water is needed to stabilize the spent shale and additional water is required for revegetation purposes. Finally, water is needed for the communities that will spring up around the shale oil plants. It has been estimated that 7,000 to 10,000 acre-feet of water will be required annually for a 50,000 barrel per day plant. Although wherever possible, the water will be cleaned and reused, the total amount of water to be supplied is tremendous, especially in a region in which the annual rainfall is less than 12 inches. The demands on existing water supplies are presently very heavy, and hence it will be difficult to obtain the water required for shale oil production.

6. Many workers will be required for large scale shale oil production. These workers and their families will need housing in this sparsely settled region. Shopping facilities and schools must be built. Police and sanitation facilities will be required. Municipal governments must be established.

If the in-situ production method is used, some of these problems, such as the mining of the shale, the disposal of the spent shale and the need of vast quantities of water, will not be critical. However, the engineering required to obtain the oil by the in-situ method is tremendous. Since this method has not been tried out in full-scale operations, it is difficult to predict its potential for shale oil production.

16.4 SUMMARY ON SHALE OIL

Particularly in the Rocky Mountain area, the United States possesses vast amounts of oil shale which has the potential of producing many times the amount of oil that can be produced from our known crude oil reserves. If our oil shale resources can be fully used, we can become independent of foreign oil and remain so for a long time. However, major problems, particularly financing, water supplies, and environmental protection, must be at least partially solved before shale oil can be produced in sufficient amounts. We will be under increasing pressure to solve these problems as the demands for oil continue to increase, as the prices rise, and as we find it necessary to import ever increasing amounts of oil. With all of the variables involved with shale oil production, it is very difficult to predict its future but it does not seem possible that significant amounts of shale oil will be produced in the next decade.

16.5 TAR SANDS

The term *tar sands* refers to a sand or a rock, particularly sandstone, impregnated with a very heavy oil, frequently known as bitumen. Unlike kerogen, bitumen is truly an oil. In some cases, tar sands resemble sand upon which used crank case oil has been poured. In other cases, the bitumen is more viscous and may exist in sandstone or other rock. The origin of tar sands is uncertain. One explanation is that as oil was formed many thousands of years ago, it had a tendency to percolate through porous rock. Upon heating, the lighter fractions of the oil are thought to have evaportated, leaving the bitumen behind. Tar sands are known also as oil sands, bituminous sands, and bituminous rock.

16.6 TAR SANDS RESOURCES

Over 500 deposits of tar sands have been identified in 22 states. It has been estimated that our tar sands resources total 33 billion barrels, with 28 billion barrels of this being in Utah. Other states containing significant amounts of tar sands are California, New Mexico, Texas and Kentucky. Estimates vary as to the amount of oil which can be recovered from our tar sands. Some estimates place this amount as high as 5 to 10 billion barrels under present conditions.

These figures do not include the so-called heavy crude. Heavy crude oil is found in sub-surface reservoirs. Apparently this bitumen collected in pools rather than impregnating sand or porous rocks. This heavy crude has too high a viscosity to be reclaimed by conventional methods. The amount of heavy crude oil is estimated to be larger than that present in tar sands.

Developmental work to obtain oil from tar sands has been initiated in Utah, California and Texas. Depending on future oil prices it is felt that one or more plants producing 50,000 barrels per day of tar sands may be in operation by the year 1995.

Although the potential for production of oil from the tar sands in this country appears to be relatively low, the reverse is true for Canada. Vast quantities of tar sands stretch over a hundred miles along the Athabasca River in Northern Alberta. Some of these tar sands lie exposed on the surface. Much is covered by overburden ranging in thickness from a few feet up to a hundred or more feet, thus presenting the opportunity for recovery by strip-mining. Still other tar sands lie buried so deep that strip mining is not feasible.

One estimate places the total amount of oil in these tar sands at 700 billion barrels. The Alberta government estimates that 26 billion barrels of low-sulfur pipe line oil can be recovered by present surface mining technology. With improved technology, it is felt that 86 billion barrels can be produced. If methods can be developed to use the deeply buried tar sands, it is felt that the total amount of oil producible from these tar sands may be as much as 300 billion barrels.

The question may be raised as to why we in this country should be concerned with the potential of Canada's tar sands. We have been importing a significant amount of oil from Canada. But the known crude oil resources of Canada are limited and to conserve its supply, Canada is decreasing the amount of oil which it sells to us and may, ultimately, terminate all oil shipments. *If* it becomes possible to produce vast amounts of oil at low prices from its tar sands, Canada may again be willing to sell us significant quantities of oil.

16.7 OIL PRODUCTION FROM TAR SANDS

Although the existence of the Canadian tar sands has been known for almost two hundred years, little effort was made to obtain oil from the tar sands until 1920. At that time, the Alberta Research Council started investigations on the use of hot water to separate the oil from the sands. Research work was continued for many years and pilot plants were erected. Finally in 1948, the Alberta government built a demonstration hot water extraction plant capable of handling 500 tons of sand per day. Although this plant demonstrated the technical soundness of hot water extraction, the economics of commercial production had yet to be demonstrated. About this time the large Leduc oil field was discovered south of Edmonton. Since crude oil could be produced at a much lower cost than was estimated for tar sands oil, interest in tar sands waned.

In the 1950's however, as demands for oil increased, efforts once again were made to develop the tar sands industry. At least one pilot plant with a capacity of 1,000 barrels of oil per day, was built. In 1967, the Great Canadian Oil Sands Ltd. put on stream the first commercial tar sands oil plant. This plant, rated at 65,000 barrels per day, apparently is producing on the average of 50,000 barrels per day. Based on the operating data obtained from this plant, a consortium, Syncrude Canada, Ltd., was organized to construct a 125,000 barrels per day plant. Syncrude is being financed by Gulf Oil-Canada, Imperial Oil Company, Canada-Cities Service, the Canadian government, the Alberta government, and the Ontario government. This plant is located on a 7,000 acre site where the overburden varies in depth from 7 to 111 feet. The tar sands deposit averages 120 feet in thickness. Much of this land consists of dry sand hills, which support only jack-pine. A large part of the remainder of the land is swampy muskeg. Only 15 percent of the area supports vegetation cover suitable for wildlife habitat.

In 1971, the estimated cost of this plant was 400 million dollars. In 1975 the cost estimate was increased to over 2 billion dollars, indicating the tremendous difficulty in making accurate cost estimates of plants much larger than those previously built. One unit of this plant came on line in 1978.

For this plant the method of producing oil from tar sands is to first strip the overburden and then to mine the sand using either bucket-wheel excavators or conventional scrappers. The sand is conveyed to the treatment plant. Here, hot water and steam are mixed with the tar sand, thus releasing about 90 percent of the bitumen, which picks up air in the process. The air saturated bitumen rises to the surface of the water as a froth. The froth is deaerated, heated and diluted with naphtha and then centrifuged. The resulting bitumen is fed into cokers to crack the molecules. In the final process, treatment with hydrogen removes most of the sulfur and upgrades the crude. The final product is a high-grade low-sulfur crude which can be piped to Edmonton, some 250 miles away, for refining. Other by-products of the coking operation, gas and coke, are used to produce the large amount of power required for plant operation.

Even though the initial cost of this plant will be many times the original estimates, it is felt that it can produce oil at a profit. Although not stated, it is probable that interest rates and taxes are lower than normal since so much of this project is financed by various government agencies. If anticipated costs can be realized in this plant, then there should be a rather large expansion in the production of oil from tar sands.

As with oil shale, consideration is being given to the in-situ method of producing oil from tar sands which are located too deep underground to use strip-mining. In most regions, it is not feasible to mine the deep veins of tar sands. Various methods have been suggested for in-situ operations. One of these methods is quite similar to that considered for oil shale, namely underground combustion with a deficiency of air. Other suggestions have been the use of hot water and steam. Various chemical means for extraction have been suggested. These include organic solvents, emulsifiers, and caustic agents. Even nuclear explosives have been suggested to heat the bitumen sufficiently to cause it to flow.

In one suggested steam system, steam is fed intermittently into the tar sands sufficiently long to heat the bitumen, thus permitting it to flow to a collection point. This method has been used in California to recover several million barrels of bitumen from sands.

In general, however, large scale in-situ operations have not been undertaken to date. Hence, it is not possible to evaluate the potential for this type of oil recovery from tar sands.

16.8 SUMMARY ON TAR SANDS

Although tar sands are found in many localities in 22 states, the amount present in most places is so small that there is little probability that it will become economcially attractive to mine it for oil. There are somewhat larger amounts of tar sands in five states. However, it appears that at the most, we will be unable to produce more oil from tar sands than we use in a year.

On the other hand, there are vast quantities of tar sands in the Athabasca River region of northern Alberta, Canada. It appears that 26 billion barrels of this oil can be readily recovered. It is predicted that it may be possible to produce a total of as much as 300 billion barrels, particularly if the in-situ method can be developed satisfactorily. It is to be hoped that if Canada can produce oil from tar sands at low costs and in large amounts, much of this oil will become available for our purchase.

SUPPLEMENT A

ILLUSTRATIVE EXAMPLES

Example 16-1. A 50,000 barrel per day shale oil plant uses shale having 25 gallons of oil per ton of shale. Determine the tons of shale to be mined per day.

Solution. The gallons per day is

$$50{,}000 \times 42 = 2{,}100{,}000 \text{ gallons}$$

The tons of shale is

$$\frac{2{,}100{,}000}{25} = 84{,}000 \text{ tons} \qquad \text{answer}$$

Example 16-2. The initial cost of a 125,000 barrel per day tar sands plant is four billion dollars. If the total fixed charges are 18 percent, calculate the fixed charges per barrel of oil produced, assuming that the plant is operated at 70 percent capacity.

Solution. The fixed charges per year are

$$4{,}000{,}000{,}000 \times 0.18 = \$720{,}000{,}000$$

The oil produced per year is

$$125{,}000 \times 365 \times 0.7 = 31{,}900{,}000$$

Fixed charges per bbl =

$$\frac{720{,}000{,}000}{31{,}900{,}000} = \$22.57 \qquad \text{answer}$$

These fixed charges alone presently exceed the total cost of domestic crude oil but are less than imported oil.

SUPPLEMENT B

PROBLEMS

16-1. The density of the shale in Example 16-1 is 155 lb. per cu. ft. Assume that the volume of the spent shale is 12 percent greater than the original shale. Determine the volume of the spent shale produced per day.

16-2. The spent shale in Problem 16-1 is to be placed in a pile 30 feet high. Determine the acres covered.

16-3. The oil shale in Problem 16-1 is obtained from a vein having an average thickness of 300 feet. How long will the oil shale in a 1,000 acre tract last?

16-4. The fixed charges in Example 16-2 can be reduced to 8½ percent by government support. Determine the fixed charges per barrel. Comment on your answer.

16-5. The lease price for a tract of oil shale is 200 million dollars. There are 30 gallons of oil per ton of shale, 80 percent of which is recoverable. A fixed cost of one dollar per barrel is allowed to pay for the lease. Determine the number of tons of oil shale which must be mined to pay for the lease.

16-6. Discuss the desirability of extensive government funding for the development of the shale oil industry in this country. If extensive funding is to be made, to what phases of shale oil production should it be devoted?

16-7. Same as Problem 16-6 except for a tar sand industry.

16-8. Assume that 1 percent of the oil which is projected to be used in 1990 is to be obtained from oil shale. Assume the heating value of the oil is 5,800,000 Btu per barrel. The shale averages 28 gallons per ton. Determine the tons of shale to be handled per year.

16-9. Using data from Problem 16-1, determine the volume of the spent shale for Problem 16-8.

16-10. Assuming that a 50,000 barrel per day shale oil plant requires 9,000 acre-feet of water per year, determine the amount of acre-feet required per year for Problem 16-8.

17

Energy from Municipal Refuse

17.1 MUNICIPAL REFUSE

All products stemming from plant or animal life contain energy and thus possess the potential for contributing to our energy supply. These products may be lumped together under the term "biomass." Biomass includes wastes produced on the farms, in the factories, and in our homes. It also includes vegetation which could be grown solely for its energy potential. In the broad sense, biomass may be considered to be a renewable energy resource, assuming that man continues the planting required for its generation. It has been estimated that the total biomass produced in this country has the potential of supplying 10 percent of our present energy needs. However, much of this biomass can be recovered only with much difficulty.

With the exception of its inert matter, municipal refuse can be classified as a biomass. Since the method of its collection and some of the methods of its disposal differ from those used for other biomass, it will be treated in a separate chapter. There is a wide variation in the composition of municipal refuse depending on how it is generated. In larger cities, much of the refuse is created by various types of industries, the nature of the refuse being dependent on the type of industry. Residential refuse and refuse from restaurants contain much food (more than they should). Paper and products constitute on the average about 53 percent of municipal refuse. Glass, metals and dirt are present in variable amounts. It has been estimated that the total municipal solid waste generated in this country amounts to about 135 million tons a year. In addition, sewage disposal systems generate approximately 14 million tons of solids per year. The energy content of these 149 million tons of solid waste totals about two Quads. Thus municipal solid waste has the potential of supplying a measurable portion of our energy needs. Unfortunately much of this municipal waste is generated in relatively small cities and towns in widely dispersed areas. It does not seem feasible to transport the waste considerable distances. Thus it appears that plants to use the energy in municipal waste will be constructed only in those areas where several hundred tons of waste per day are available locally.

In the past, municipal refuse has been treated primarily as a source of problems of collection and disposal, with little thought being given in this country to its energy potential. Years ago, some municipal refuse was dumped out in the country or, when produced near the ocean, barged out to sea and dumped, with perhaps some of it washing ashore later. Much of the refuse was burned in open city dumps. Later, crude and polluting incinerators were used extensively for refuse disposal. Some attempt has been made to reduce the refuse to a useful product by composting it.

Today, under pressure from the environmentalists, most of the municipal refuse is disposed of in sanitary landfills. The concept here is to dump the refuse on more or less useless land, covering it with dirt as it is dumped. When properly done, this method has not only disposed of the refuse but it has created useful land. However, the landfill method has not been fully satisfactory. In some cases, water has leached harmful substances from the landfills, and has caused pollution. In many of the landfills, after a period of time, methane gas has been produced. Some explosions have resulted in buildings constructed over the landfills. Some cities, such as Los Angeles, are attempting to collect the methane and use it. As cities grow and as the more readily available landfill sites are used up, it becomes necessary to use land far out in the country. Not only has this land become expensive but also the costs of hauling the refuse long distances has made its disposal very costly. These factors, together with the realization that refuse has the potential of substantially increasing our supply of energy, are a motivation for developing methods of using this energy and at the same time minimizing the disposal problem. Although other methods are being considered for municipal refuse disposal, most of the present effort is being directed toward either incineration or pyrolysis.

17.2 INCINERATION

The early incinerators used for municipal refuse disposal often were little more than enclosures to confine the burning refuse with grates to permit air from beneath. Most of these incinerators were unsightly, odor causing, and polluting. As such, incinerators acquired a poor image which still persists in the eyes of many. The primary objective of the early incinerators was to greatly reduce the volume of the refuse as well as to produce an inert product for easy disposal. Little or no thought was given to energy use.

For many years, incinerators have been used in large European cities to produce steam, generally for electric power generation. Over the years in this country, there have been isolated installations of modern incinerators. The nature of the incinerator and the method of energy use is governed by the nature of the refuse supplied to it. As stated earlier, paper constitutes a large portion of municipal refuse and contributes much to its heating value. Some effort is being made to separate the paper from the refuse and reclaim it. If this is done, the heating value of the refuse and hence its potential for supplying significant amounts of energy are materially reduced. However, much of the paper is not readily separable and the incentive for attempting to do so is reduced if its energy can be recovered by either incineration or pyrolysis. Thus, it appears, in general, that if incinerators are installed for energy production, paper will not be separated from the rest of the refuse.

It is necessary to reduce the refuse to uniform small particles in order for the incinerator to operate in a substantially pollution-free manner. The common method of making this size reduction is to use hammermills as a shredding process. In some cases, the hammermill may reduce the refuse to a maximum size of 8 inches (i.e., will pass through an 8-inch by 8-inch opening). The refuse then is fed directly into the incinerator. In other cases, particularly when the process to be used is pyrolysis instead of incineration, the refuse is passed through a second shredder. The particle size leaving the secondary shredder varies greatly depending on the type of installation but may be as small as 0.015 inches.

After the shredding, metals and glass may be separated from the combustibles. Large magnets, when passed over the shredded refuse, will remove ferrous metals. An air classifier is used to separate the combustibles from the remaining inert materials. In the classifier, air is forced up through the refuse with sufficient velocity to carry the light combustibles with it. The heavier inert materials fall to the bottom. It is possible to separate the glass from the rest of the inert materials and then reprocess it.

The separated glass, iron and aluminium, can be sold for reprocessing, thus providing some revenue. The solids remaining after combustion are inert and have a volume much less than that of the original refuse, making its disposal for land fill purposes is not much of a problem.

The refuse is forced into the incinerator and moves along on some type of grate. The major portion of the air, the primary combustion air, is fed in beneath the grate with the remainder, the secondary air, being supplied to the combustion chamber above the grate. Not only must this secondary air be supplied in the right amount, but it must be so directed that the combustion will be substantially complete. Finally, some type of a gas cleaning device must be installed to prevent any significant amount of particulates, as well as polluting gases, being discharged into the surroundings.

The most obvious way to use the energy of municipal refuse in connection with an incinerator is to produce steam for driving a steam turbo-generator unit for electric power generation. As stated earlier, this is being done in several European cities and in a few cities in this country. There are, however, several problems associated with this method of power production. Zoning laws permitting, an incinerator can be located relatively close to the center of refuse production, thus saving the very large costs now incurred in trucking the refuse way out to suitable landfills. However, the large amount of water required for the power plant condensers frequently is not available in central locations. Although electric power production from refuse may be justified economically for the largest cities, this is not true for smaller cities. For an average city of 100,000, the refuse produced may be sufficient to generate 7,000 kW continuously. The initial cost per kW rating is very large for a small plant of this size. In addition, the labor costs for operating this size plant may be almost as large as those in a much larger plant. Although it is not economically feasible for a medium size city to produce power from municipal refuse in a steam power plant built for this purpose, power can be produced from refuse when a public utility power plant is presently located reasonably close to the center of refuse generation. If the plant is coal fired, refuse can be substituted rather easily for up to 10 percent of the coal being used.

Such an arrangement has been used for several years in St. Louis, Missouri. Refuse is fired in two 20-year-old boilers, each designed to provide steam for 125 MW turbo-generator units. The boilers were designed to operate on either pulverized coal or natural gas. Refuse supplies about 10 percent of the heating requirements and thus saves about 300 tons of coal per day. For the year 1972, the preparation costs chargeable to the city of St. Louis was $5.00 per ton of refuse. The revenue realized from the reclaimed metal was $1.00 per ton of refuse, making the net costs to the city of $4.00 per ton of refuse. Figures are not available for what the costs would have been to the city had the refuse been disposed of in sanitary landfills. Normally, it costs cities much more than $4.00 per ton for landfill disposal. The preparation cost to the Union Electric Company was $1.05 per ton of refuse burned for which a saving of fuel was made of $4.20. Thus, the power company made a profit of $3.15 per ton of refuse burned. Unfortunately, only a few cities have steam power plants located as conveniently as does St. Louis. For larger cities, even without conveniently located existing power plants, it is recommended that a careful investigation be made of the feasibility of electric power generation using municipal refuse as the energy source, particularly if it is possible to tie in electrically to the local power system.

An alternative to electric power production from municipal refuse is the production of steam which could be sold to local industries. Steam is a very satisfactory medium for supplying heat at moderate temperatures, say up to at least 400°F, since its temperature remains constant as it condenses, provided its pressure is held constant. Because of this fact, many industries generate steam for the heating required in their various processes. They also may use steam for building heating. In some cases, industries needing appreciable quantities of steam are located or can be located near a desirable site for a steam-producing municipal incinerating plant. Under these conditions, it is conceivable that the municipal plant could produce steam at a lower cost than that of disposal of its refuse by landfill use. The steam might be sold to industries at a cost lower than that of producing it by the industries. Certainly this possibility should be investigated fully in any long range planning.

17.3 PYROLYSIS

Pyrolysis of refuse is described as the method of heating the refuse in the absence of air. Temperatures as high as 1,800°F are used. This heating not only drives off the volatile constituents of the refuse but also decomposes the less volatile organic materials. The products produced from the organic part of the refuse are a low Btu gas, oil, and char. The oil produced is generally tar-like, high in oxygen, and with a low heating value. The heating value of the oil is dependent on the nature of the pyrolysis process as well as on the composition of the refuse. In general, it ranges from 10,500 to 15,000 Btu per pound, compared with 18,200 Btu per pound for No. 6 fuel oil. Since most municipal refuse is low in sulfur, the sulfur in the gas and oil produced by pyrolysis is not environmentally excessive.

As with incineration, the refuse is shredded before being fed into the pyrolysis reactor. In some cases, a secondary shredder reduces the refuse to a maximum size of an eighth of an inch. Glass, metals and other inert solids in the refuse may be removed, as in the case of incineration, before entrance to the reactor or they may be separated from the pyrolysis residue.

Several groups in this country are working on the pyrolysis method of utilizing municipal refuse. The three most advanced groups are the Monsanto Enviro-Chem Systems, the Garrett Corporation, and the U.S. Bureau of Mines. Each of these groups has operated pilot plants and larger ones are planned or are in operation. A 1,000 ton per day Monsanto Landguard plant is now operational in Baltimore, Maryland. The gas produced has such a low heating value (100 Btu per standard cubic foot) that it is not feasible to transport it for use elsewhere. For this reason, it is burned where produced, generating 4.8 million pounds of steam per day. This steam is sold to the Baltimore Gas and Electric Company for use in its steam distribution system.

The objective of the Garrett process is to maximize the production of oil. This is accomplished by drying the refuse, highly shredding it and by keeping the reactor temperature relatively low. This oil, known as garboil, has chemical characteristics differing significantly from those of crude oil and hence cannot be readily refined into the customary petroleum products. However, it can be readily burned. As of 1975, a 200 ton per day plant was under construction to handle the refuse of Escondido and San Marco, California, with the oil to be sold to the San Diego Gas and Electric Company.

Unlike the Monsanto and Garrett processes, which are designed for full resource recovery, the present U.S. Bureau of Mines effort is being directed more to a study of the pyrolysis reaction taking place in a pilot reactor, into which various substances are charged. The reactor basically is an electrically heated, cylindrical chamber.

The efficiency (i.e., the ratio of the heating value of the products to that of the original charge) ranges from less than 50 percent to over 70 percent for those pyrolysis systems which have been built. Since operating data is not available from full-size plants, the economics of the pyrolysis method of municipal refuse disposal cannot be compared with those of incineration and sanitary landfill. It is true that the City of Baltimore decided that it was economically sound to build a large pyrolysis plant. However, much of the money to construct this plant came from the Environmental Protection Agency. Present indications are that, in general, it probably will become economically sound to use the pyrolysis method for municipal refuse disposal. This may be particularly true when sizeable quantities of oil are produced, which can be readily transported and sold.

17.4 SUMMARY OF MUNICIPAL REFUSE

It has been estimated that the energy in the readily collectible municipal refuse is almost one percent of our total present energy requirements. In itself this energy is not very significant. However, when combined with similarly small amounts of energy which can be obtained from the other alternative sources and by conservation efforts, the total becomes significant.

Most of the efforts towards using the energy of municipal wastes have involved incineration, with some efforts being devoted to the use of pyrolysis. Sufficient information is not available from full-scale plants to judge the economic feasibility of either of these disposal methods. In general, however, it appears that the use of incineration to produce steam for electric power generation may be desirable for the very large cities, particularly if it is possible to integrate the operation with that of a local power company. When steam produced by incineration can be sold to local industries, it may be also advantageous to use incineration in somewhat smaller cities. Particularly where oil is a major product of a pyrolysis operation, it may be desirable to use this method for refuse disposal for substantially smaller cities.

The funds for most of the existing incineration and pyrolysis plants for municipal wastes have come, to a large extent, from government sources, federal, state, and municipal. In general, the existing plants have not demonstrated the economic feasbility of treating municipal waste without governmental financial aid. As might be expected, several of these plants have encountered operating difficulties. Thus, until more operational data become available, it will be difficult to obtain, from private investors, the large sums of money required to construct plants to recover energy from a large portion of our municipal waste.

Not only will the initial costs of incineration and pyrolysis plants be high but the operation costs also will be high. Balanced against these costs will be the revenue produced by the plants and the savings made in landfill operations, not only in the cost and operation of the landfill itself but also the cost in trucking the waste long distances to the disposal sites. However, efforts should be continued to use the energy in municipal waste because of its potential for making a contribution to our total energy supply and for minimizing the ever growing problem of municipal waste disposal.

SUPPLEMENT A

ILLUSTRATIVE EXAMPLES

Example 17-1. An incineration plant is being considered for a city of 150,000 population. Assuming an overall thermal efficiency of 25 percent and using average figures from the text, estimate the average amount of electric power which can be produced.

Solution. Take the average per capita daily production of refuse to be 6.5 pounds and its heating value 5,200 Btu per pound. The average amount of refuse produced per hour is

$$\frac{150{,}000 \times 6.5}{24} = 40{,}630 \text{ lb. per hr.}$$

The power output equals

$$\frac{40{,}630 \times 5{,}200 \times 0.25}{3{,}413} = 15{,}470 \text{ kW} \qquad \text{answer}$$

Example 17-2. Assume that paper and paper products constitute 53 percent of its refuse in Example 17-1 and account for 71 percent of the heating value. Determine the average amount of power which can be produced from the refuse if 75 percent of the paper is removed for recycling.

Solution. The heating value of the paper in the refuse is

$$0.71(5{,}200) = 3{,}692 \text{ Btu/lb. of refuse}$$

The heating value of the paper removed is

$$0.75(3{,}692) = 2{,}770 \text{ Btu/lb. of refuse}$$

The heating value of the remaining refuse is

$$5{,}200 - 2{,}770 = 2{,}430 \text{ Btu/lb. refuse}$$

The power output is

$$\frac{40{,}630 \times 2{,}430 \times 0.25}{3{,}413} = 7{,}230 \text{ kW} \qquad \text{answer}$$

SUPPLEMENT B

PROBLEMS

17-1. Assume that 85 percent of the energy of the refuse in Example 17-1 which is not delivered as electrical energy is given up in the condenser.

Assume that the temperature rise of the circulating water in the condenser is 15°F. Determine the pounds of circulating water passing through the condenser per second. Comment on your answer.

17-2. Why is the thermal efficiency of the incinerator plant in Example 17-1 so much lower than that normally associated with public utility power plants?

17-3. Outside of its appearance should there be any objection to the location of either an incinerator plant or a pyrolysis plant in a place central to the generation of refuse? Why?

17-4. In deciding to install an incinerator plant for a large city, would you recommend one very large efficient unit or two or more smaller, less efficient units? Why?

17-5. Based on the answers obtained in Example 17-1 and 17-2, comment on the desirability of using incineration for refuse disposal and power production if it is feasible to remove most of the paper from the refuse.

17-6. Determine the average electrical power output from the readily collectible municipal refuse of 71,000,000 tons per year if the average thermal efficiency of the plants is 25 percent. Compare your answer to the average electric power now being produced.

17-7. A small city (population 15,000) now trucks its municipal waste about 4 miles (average) to a sanitary landfill. Discuss the desirability of

a. Installing an incineration plant to generate electrical energy.
b. Trucking the waste 14 miles to a larger city having an incinerator plant.

17-8. In case it is not desirable to use incineration for Problem 17-7, should both residents and industry be required to separate paper and paper products from the rest of their waste?

18

Energy from Biomass

18.1 BIOMASS RESOURCES

The biomass present in municipal refuse was discussed separately in Chapter 17. In this chapter, we will consider the other biomasses which have a potential for adding to our energy supply. Principally, these include wood and wood waste, animal waste, vegetation and vegetation waste, and sewage.

It is extremely difficult to evaluate the total amount of biomass produced in this country, since most of it becomes waste and largely is discarded. As a result, there is a wide range in the estimates of the amount of these biomasses. Not including municipal refuse and wood used for wood products, perhaps 750 million tons of dry, ash-free biomass are produced annually in this country. However, the amount of this biomass which is readily collectible is probably less than 10 percent of the total. The heating value of this readily recoverable biomass is equivalent to almost 100 million barrels of oil. Presently, no significant portion of this energy is being recovered. The broad problems relative to biomass then are how to use the readily collectible biomass and how to justifiably collect much more of it.

18.2 WOOD

In our early history, wood was not only our principal building material but was also our chief source of energy. Gradually, with the very large growth in our demands for both energy and for wood and wood products, our forests were unable to meet all these demands. With the use of first coal, then oil and natural gas, fossil fuels took the place of wood. Today, wood, largely wood wastes, supplies only about one percent of our energy needs. Because wood is a renewable resource, it is natural to consider the possibility of wood once again supplying a significant portion of our energy.

Our total forest area is still very large, approximately 754 million acres or one-third of our total land area. This area is about three-quarters of the area occupied by forests when Columbus landed here. Almost one-third of our forest area is set aside for parks, wilderness areas, watersheds, or is not suitable for growing commercial timber. This leaves 500 million acres of commercial forest. Of this 500 million acres, 27 percent is publicly held, 60 percent privately owned by about 4 million persons, and the remaining 13 percent owned by the forest products industry. About 26 percent of the wood produced in this country comes from the land owned by the forest products industry. This fact illustrates the possibility of greatly increasing our total wood production, particularly on the government owned land. It is possible to greatly increase our wood production by applying well-developed forest management technology.

Our total commercial forest area is decreasing slowly as some is withdrawn for wilderness and park uses and also to permit urban growth. At the same time, our demand for wood and wood products is increasing significantly. It has been estimated that our demands for forest products in the year 2000 will be double our present demands. Even at present, we find it necessary to import much of our pulp and paper products, particularly from Canada. It may be possible to meet our increased needs for wood products and still have some wood available as an energy source if we can maximize production on the 87 percent of the commercial forests which are government owned or privately owned.

Wood has the potential of supplying energy for power generation and for heating. Much wood for these purposes can be obtained from the residue after commercial lumber is removed from wood lots. Some estimates place the amount of energy in the wood left behind about equal to that in the timber removed. In the past, little effort was made to salvage the limbs and other woody products left behind. Often it was burned in open fires to permit reforestation. Today, with increasing costs of conventional fuels, a considerable effort is being made to salvage this waste wood. In some cases it is chipped and may be pelletized to aid in transportation and in burning.

A considerable portion of forest lands do not produce commercial timber. The trees grown on these lands have the potential for producing a sizeable amount of energy. Thus the total potential of wood energy is significant.

There is some feeling that wood-burning power plants can provide an appreciable amount of electric power. It should be recognized, however, that vast areas of forest are required to supply continuously the wood required for a large power plant. For a 1,000,000 kW power plant this area may be around 1,500 square miles, depending on the type of trees and the growing conditions. This is an area of about one and a half times that of the state of Rhode Island.

For many years, a considerable amount of wood has been burned in fireplaces. The ordinary fireplace is very inefficient in supplying heat to a room. Furthermore, unless the fireplace damper is closed when there is no fire, much of room air, heated by other means, will go up the chimney. Recently, though, a considerable effort has been devoted to improving fireplace efficiency. A common method of doing this is to circulate room air through tubes placed in the fireplace. Although this does make a significant improvement, the efficiency is still low. A second development in the burning of wood has been the reintroduction of wood burning stoves. Some of these stoves are relatively efficient and to a limited extent are replacing conventional residential heating systems. This is an economical method of heating when wood is available on one's own property. On the other hand, when wood must be purchased, particularly when it must be hauled for some distance, this method of heating may be more expensive than conventional methods.

There is a growing use of wood furnaces in large buildings and in many industries. In general, either wood chips or wood pellets are burned. When this wood can be obtained from residues when timber is harvested, it may be more ecnomically desirable than to burn oil. The same statement may be true for wood obtained from trees which will not produce commercial timber. Wood wastes are also being used increasingly by wood industries, particularly furniture manufacturers.

There are some proposals for establishing large tree farms to produce wood for power production. The advocates of these proposals feel that, together with wood burned for heating, wood burned for power helps reduce dependence on other fuels. It is very difficult to predict the percentage of our energy needs which can be met by wood. Several factors should be considered in predicting the possible growth in the use of wood energy:

1. Trees, as well as other vegetation, are very inefficient in the use of solar energy. The average efficiency for trees is around 0.2 to 0.6 percent (i.e., between 0.2 and 0.6 percent of the solar energy received by a given area in a year is converted into the energy of the wood in the trees on that area).
2. Vast areas will be required to produce the desired amount of wood. Where will this land come from? With the heavy, increasing demands for food, can we give up a large portion of our farm lands? With the heavy demands for wood products, can we convert our timber producing lands to producing wood for energy uses?
3. It is true we have a considerable amount of land in this country which is sub-marginal, but much of this land is sub-marginal because of a lack of water, making it unfit for timber production.
4. In many of these areas, fertilizers must be used to produce the desired growth, thus adding to the cost of wood. Furthermore, energy is required to produce the fertilizers.

It is true, as indicated above, that wood yields can be increased by the use of proven forest management techniques. Furthermore the efficiency of solar energy utilization can be increased by special breeding techniques. In certain parts of the country having large forest areas, wood energy can make a significant contribution to our energy supply. For the country as a whole, it does not appear that wood can ever make a large contribution to the energy supply; however, efforts should be made to use as much wood energy as possible to help meet our energy demands.

18.3 VEGETATION

Since wood has already been discussed in Section 18.2, it will not be included in this section on vegetation. Most of the vegatation grown in this country is grown for food for humans and animals, or for fiber. As such, varieties have been developed to produce the highest yield, consistent with a product of high quality. Under these conditions, the efficiency of solar energy use by the entire plant is not good. For some plants this efficiency is lower than that of trees, being as low as 0.1 percent. Some other plants, such as sugar cane, may have a solar energy conversion efficiency as high as 1.2 percent. If plants were to be developed solely for that purpose, it should be possible to attain higher energy conversion efficiencies. This has led to the consideration of developing and growing plants solely for their energy producing potential. In general, the same four factors that were identified in Section 18.2 for the production of power from wood apply equally well to the production of power from other vegetation grown for that purpose. Because of the need for retaining our farm lands for food and fiber production, it does not appear that a significant amount of vegetation will be produced solely for its energy content.

On the other hand, the amount of agricultural crop wastes is significant, the amount generated annually being in the order of 400 million tons on the dry basis. At present only about 5 to 6 percent of this crop waste is readily recoverable. No significant portion of it presently is being used for its energy content. Because of its potential, various methods have been suggested for using this energy. Crop waste can be dried (perhaps field dried), shredded and burned in incinerators in a manner similar to that used for municipal refuse. Some pilot plant work is under way to use hydrogeneration for energy recovery both from municipal refuse and crop wastes. In this process, the waste, together with a catalyst (such as sodium carbonate) are placed in a reactor. Carbon monoxide and steam are introduced at pressures ranging from 100 to 250 atmospheres. The retort is heated to temperatures ranging from 500°F to 750°F. The residence time in the reactor ranges from one to two hours. Apparently, the carbon monoxide and steam react to form carbon dioxide and hydrogen. The hydrogen combines with the organic material to form a heavy oil, relatively high in both oxygen and nitrogen. Its heating value is about 83 percent of that of No. 6 oil. It has been estimated that, on the average, a ton of dry waste will produce about two barrels of oil, with three quarters of a barrel of it being used to furnish the heat required in the process. Although this technology for producing oil from organic wastes has been demonstrated, the economic feasibility cannot be determined until large plants are built. One disadvantage of this system is the high pressure required in the retort.

Much of the crop wastes in fields is plowed under to enrich and condition the soil. The value of the crop wastes for this purpose must be balanced against their energy potential when they are removed from the fields.

18.4 GASOHOL

Recently, much interest has been developing in the use of gasohol in automotive engines. Commonly, gasohol consists of 90 percent gasoline and 10 percent ethanol (ethyl alcohol). Because ethanol is made by fermentation of grain (a renewable source) or similar cellular substances, its use could reduce our need for imported oil. Apparently, when the ethanol is pure, no significant problems have developed in the use of gasohol. However, several factors must be considered in judging the extent of the future use of gasohol:

1. Operating difficulties may be encountered if the ethanol contains significant amounts of water.
2. The heating value of ethanol, on the gallon basis, is about two-thirds that of gasoline. Hence, other things being equal, somewhat fewer miles per gallon will be obtained with gasohol than with gasoline.
3. Energy must be used to produce fertilizers for growing the grain, harvesting it, and making the ethanol. The energy in the ethanol is significantly less, perhaps one half of that of the grain from which it is produced in large units. In small units more total energy may be required to produce ethanol than its energy content.
4. Using grain for producing ethanol decreases the supply which we can sell to hungry people throughout the world. The income from foreign sale of grain helps to decrease our unfavorable trade balance.

Two positive factors for the use of gasohol are

1. Because ethanol resists engine knocking better than untreated gasoline, gasohol can replace high test gasoline for use in higher compression ratio engines.
2. Because ethanol is produced from a renewable source, our dwindling supplies of gasoline are conserved.

Although there is an increasing use of gasohol, it is difficult to predict whether or not gasohol will significantly lower our future demand for gasoline.

There is some effort to produce methanol, (methyl alcohol) from wood or wood and other wastes, to be blended with gasoline in a manner similar to that of ethanol. Although research with this use of methanol has not progressed as far as with ethanol, the factors relative to ethanol seem to apply in general to methanol.

18.5 OTHER BIOMASS SOURCES

Two other biomass sources which are being considered for their energy potential are animal wastes, particularly manure, and sewage. It has been estimated that some 200 million tons of manure, dry basis, are generated per year in this country. Of this amount perhaps somewhat over 12 percent is readily collectible. Much of this readily collectible manure is generated at the feed lots where the animals are fattened before being shipped for slaughter.

The potential of energy obtainable from sewage is relatively small. Approximately 12 million tons, dry basis, of sewage solids are generated per year. Of this amount only 10 to 15 percent is readily collectible.

The method receiving the most attention for using the energy in both manure and sewage is that of producing methane by the anaerobic degradation, using microorganisms inherent in the wastes. The main purpose in sewage treatment is to separate the water from the solids and to purify the water sufficiently that it can be discharged into streams or rivers. Much of the solids is insoluble organic matter which is decomposed by an anaerobic process, involving a complex series of digestive and fermentative reactions. The products of this treatment are methane, carbon dioxide and sludge. The sludge, when dried and stabilized, is odorless and is a good soil conditioner. It can be used directly for this purpose or it can be sterilized, pelletized and bagged to be sold as fertilizer. When an incinerator plant is located close by, it may be desirable to burn the dried sludge for its energy content.

In a few of the larger sewage treatment plants, the methane produced is collected and used in the plants themselves, largely as a source of heat. In the other plants, it is wasted. In general, the operating conditions in the plants, particularly the temperatures, are controlled to give optimum performance in the treatment process. A larger amount of methane can be produced by adjusting the operating conditions for this purpose. However, in general, the residence time will be somewhat longer and hence the amount of sewage handled by a given plant may be reduced. Because of the growing shortage of natural gas and the anticipated increase in its price, a careful examination should be made of the economic feasibility of maximizing the production and use of methane from sewage.

This anaerobic process for methane production from manure has been demonstrated to be technically feasible in several small pilot plants. Because of the high rate of production of manure in the larger animal feed lots, the potential exists for generating significant amounts of methane. The economics of this production of methane cannot be established until operating data are obtained for full-size plants. In evaluating the desirability of producing methane from manure, consideration should be given to the alternative value of the manure as a fertilizer and a soil conditioner.

Three other biomasses are being considered for their energy potential. These are algae, phytoplankton, and kelp. Algae may be described as a group of plants having no true roots, stems, or leaves. Algae are found in water or in damp places. The conversion efficiency of solar energy by algae is much higher than the average conversion efficiency of 0.4 to 1.2 percent for conventional plants. Conversion efficiencies for algae have been reported as ranging between 4 and 10 percent outdoors and as high as 20 percent for carefully controlled laboratory studies. Plankton is described as microscopic animal or plant life found floating or drifting in the oceans or in bodies of fresh water. Both algae and plankton require, in addition to solar energy, a supply of nutrients. The cold water from the depths of the ocean contains much more of the desired nutrients than do the surface waters. An experimental mariculture farm is now in operation off the coast of St. Croix in the Virgin Islands, in which the nutrients are obtained by pumping water from over a half mile below the surface of the ocean. It has thus been demonstrated that it is technically feasible to grow food for fish and other marine animal life. Based on these results, it is proposed to establish mariculture farms in connectin with thermal-gradient power plants. The circulating water leaving the condensers contains the desired nutrients since it was pumped from the ocean depths. The production of fish life in mariculture farms is not a direct contribution to our energy supply but it could free some of our farm area for other uses and save the energy which was required to farm the land.

In addition to the production of fish, it has been suggested that algae be grown for its direct energy possibilities. The use of the energy potential of both algae and phytoplankton has not been carried out on a sufficiently large scale to determine its economic feasibility. However, because of the energy potential which is known to exist, it is desirable to make further investigations.

Kelp is a fast growing seaweed found in many parts of the oceans. It has been harvested to a limited extent for the chemicals which may be reclaimed from it. Some thought has been given to the development of kelp farms. It is felt that the kelp could be grown even more rapidly under controlled conditions. The kelp could be raised both for its chemicals and for its energy potential. Since so little work has been done in this area, it is not possible to evaluate the potential for kelp supplying us with a significant amount of energy. However, further investigation is desirable.

18.6 SUMMARY ON BIOMASS ENERGY

In addition to the energy in municipal refuse, the total energy of biomasses is very large. If all of this energy could be used, it would make a very significant contribution to our energy supply. However, only a small percentage of these biomasses are sufficiently concentrated in specific areas to make their use at all feasible. Wood years ago supplied a large portion of energy needs. However, as a result of greatly increased demands for wood products, wood and wood wastes now supply only about one percent of our energy needs. With our ever-increasing demands for food production and for wood products, it does not seem probable that we can expect very much wood will be grown for its energy potential. Possible exceptions are an increasing use of wood in fireplaces, in wood stoves, and in heating furnaces, and perhaps also in some power generation in restricted areas where sizeable quantities of wood are grown which cannot be made readily into wood products. In fact, unless we greatly increase our yields, there may be a severe shortage of wood for wood products by the year 2000.

The efficiency of conversion of solar energy by vegetation, including forests, is very low, generally much less than one percent. Hence, it does not seem probable that any significant amount of vegetation will be grown for its energy potential. The possible exceptions are algae and kelp, whose energy conversion efficiency is substantially higher than that of other vegetation. However, the economic feasibility of growing these vegetations for their energy potential has not been established. On the other hand, wood wastes make some contribution to our energy supply and have the potential for a larger contribution. Although not presently used, the wastes from other vegetation could supply a small portion of our energy. Investigational work is required to determine the economic feasibility of using this potential energy source.

Gasohol, containing 10 percent ethanol produced from grain, is becoming popular for automotive fuel. Several factors are involved in determining the economic feasibility of extensive use of gasohol.

Both sewage and manure present the potential for producing sizeable quantities of methane. Sewage treatment plants now produce some methane but could produce more if the economics of doing so were attractive. Information is lacking to establish the economic desirability of producing methane from manure.

Since the total energy in our various biomasses is large, they have the potential for satisfying a significant amount of our energy needs. In general, operating data from large plants will be necessary before the extent of the use of biomass energy can be predicted. However, because the potential is large, much effort should be used to develop these energy sources.

SUPPLEMENT A

ILLUSTRATIVE EXAMPLES

Example 18-1. The text states that a 1,000,000 kW power plant, operating under normal conditions, would require a forest area of approximately 1,500 square miles to supply the wood continuously for the plant. Calculate the corresponding efficiency of utilization of solar energy by the trees. Assume a plant capacity factor of 0.7, a power plant efficiency of 36 percent and an average value of solar insolation of 1,400 Btu per sq. ft. per day.

Solution. The total area is

$$1{,}500 \times 640 \times 43{,}560 = 4.18 \times 10^{10} \text{ sq. ft.}$$

The heat received per hour equals

$$\frac{4.18 \times 10^{10} \times 1{,}400}{24} = 2.44 \times 10^{12} \text{ Btu}$$

The heat supplied to the plant equals

$$\frac{\text{Output}}{\text{Efficiency}}=\frac{1{,}000{,}000\times 0.7\times 3{,}413}{0.36}=6.64\times 10^{9}\text{ Btu/hr}$$

The efficiency of solar energy utilization equals

$$\frac{6.64\times 10^{9}}{2.44\times 10^{12}}=0.00272\text{ or }0.272\text{ percent}$$ answer

SUPPLEMENT B

PROBLEMS

18-1. It is proposed to use a 3,000 acre area near Atlanta, Georgia, for a tree farm in connection with power generation. Assuming that the efficiency of the trees in utilizing solar energy is 0.5 percent and the average power plant efficiency is 34 percent, estimate the average plant output in kW.

18-2. It has been suggested that we could save a significant amount of fossil fuel by burning wood in fireplaces. Assume that wood has a heating value of 20.5 million Btu per cord and that the fireplace efficiency can be improved to 10 percent. How many cords of wood must be burned per day to reduce our energy demands by one half of one percent in 1990? See Table 2-1 for our projected total energy demands at that time.

18-3. Using data from Problem 18-1, estimate the total acreage required if we were to grow wood to produce 5 percent of the electrical energy contemplated for the year 1990.

18-4. It has been estimated that 2,000 scf (standard cubic feet) of methane can be produced by anaerobic digestion per ton of dry, solid wastes. The heating value of the methane is 1,000 Btu per scf, approximately equal to that of natural gas. Calculate the tons of solid waste required per day to produce the equivalent of one percent of the natural gas which is anticipated will be used in 1990. See Table 2-1.

18-5. Considering your results from Problems 18-1 through 18-4, discuss the probability of wood and solid wastes supplying a significant portion of our energy needs either in the near future or in two or three decades from now.

18-6. Discuss the desirability of extensive research and development being expended on each of the various biomass energy sources described in this chapter.

18-7. Discuss the desirability of greatly increasing the use of wood for heating.

18-8. Is it reasonable to expect equal increases in the use of biomass energy in all parts of the country? Why?

19

Nuclear Fusion

19.1 INTRODUCTION

Much has been written lately, particularly in the press, about the potential of nuclear fusion. It is recognized that the tremendous energy generated within the sun comes from the fusion of hydrogen atoms. Since hydrogen can be obtained from water, almost unlimited supplies of hydrogen are available in the oceans of the world. Thus it appears that nuclear fusion of hydrogen could supply us with all the energy we demand and could continue to do so for ever. Consequently, it is felt by many that we should concentrate all of our efforts on perfecting the fusion process rather than trying to use other energy sources which have limited potential. It is stated that we now have the technology for using the energy of nuclear fusion, as evidenced by the hydrogen bomb. Hence, there is some feeling that the selfish interest of the larger energy companies must be blocking the development of nuclear fusion.

It is the purpose of this chapter to examine the fusion process and its present development as well as its potential.

19.2 THE FUSION PROCESS

As stated in Chapter 9, an atom consists of a nucleus surrounded by electrons which move in orbits around the nucleus. The nucleus consists of protons and, with the exception of ordinary hydrogen atoms, neutrons. Fission can take place when the nucleus of certain heavy atoms are bombarded with neutrons, thus splitting the atom. During the fission process, some of the mass of the nucleus is converted into energy. Fusion, on the other hand, occurs when the nuclei of two atoms are joined together. In this process also, there is transformation of mass into energy.

Since the proton is a positively charged particle, the nucleus is also positively charged. But like-charged particles repel each other. To overcome the high repulsive force existing between the nuclei, the fusing nuclei must be given very high kinetic energies. Since the repulsive force is proportional to the mass of the nucleus the energy required for the fusion process is also proportional to the mass of the nucleus. For this reason only the very lightest of the atoms, namely hydrogen, is being considered for the fusion process.

It is possible to impart the desired amount of kinetic energy to the nuclei by placing them in a charged-particle accelerator, such as a cyclotron. This has been done in laboratory experiments using deutrons (the nuclei of the hydrogen atom isotope deuterium – consisting of a neutron and a proton). The accelerated deutrons are directed onto a target containing other deutrons. Although a very small amount of fusion will occur, most of the high velocity deutrons are scattered and lose their kinetic energies, which goes to heating the target. The target also has a tendency to absorb the energy released by fusion. A more practical way to give the nuclei the desired amount of kinetic energy is to heat them up to exceedingly high temperatures, perhaps 100,000,000°K. This required temperature is dependent on the type of nuclei involved. Hydrogen has three isotopes, ordinary hydrogen having one proton but without a neutron, deuterium having one proton and one neutron, and tritium having one proton and two neutrons. It appears that the fusion taking place in the sun involves ordinary hydrogen nuclei. This reaction rate is slow, much too slow to be considered for power generation on earth. In the sun, however, there is a tremendous amount of hydrogen present. Thus, even though only a very small percentage of the hydrogen nuclei fuse together in a given time, the very large number of nuclei present makes possible the very large amount of energy produced.

It should be noted here that when hydrogen is heated to very high temperatures, it becomes ionized, with the electrons being driven off and the positively charged nuclei left behind. The combination of the free electrons and positively charged nuclei constitute what is called a plasma.* A knowledge of plasma physics is necessary to understand fully the behavior of the plasma in the production of fusion energy. For our purposes at present, it is necessary for us to realize that the nuclei move with random velocities and that there is a wide range in the velocities of the various nuclei. Only relatively few nuclei possess sufficient kinetic energy to overcome the repulsive forces existing between the various nuclei and thus enter into the fusion process. Of course, the higher the temperature, the larger the number of nuclei which possess the desired amount of kinetic energy. The number of high velocity nuclei formed from the ordinary hydrogen is much too small to allow the fusion process to be sustained, even at temperatures as high as 100,000,000°K.

Deuterium is being seriously considered for its fusion potential. A deutron, the nucleus of a deuterium atom, is composed of a neutron and one proton.

* In general, a plasma consists of electrons and positive ions. For most substances under most conditions, the ions are composed of the nucleus together with those electrons which have not been freed from the atom. In the case of hydrogen at very high temperatures, substantially all of the electrons have been freed and hence the ion consists only of the nucleus.

For approximately 6,500 atoms of ordinary hydrogen present in water, there is one atom of deuterium. Thus 30,000 pounds of water contain approximately one pound of deuterium. The deuterium present in one gallon of water has the potential of producing the same amount of energy as is present in 300 gallons of gasoline. Because of the very large amount of water present on the face of the earth, it should be evident that the potential of energy production from deuterium is almost beyond comprehension. The cost of producing deuterium from water is relatively low.

There are two possible fusion reactions between deutrons (D):

$$D + D \rightarrow He^3 + n + 3.2\,\text{Mev}^* \qquad (19\text{-}1)$$

$$D + D \rightarrow T + H + 4.0\,\text{Mev}^* \qquad (19\text{-}2)$$

He^3 is an isotope of helium, composed of one neutron and two protons
n is a neutron
T is the nucleus of a tritium atom consisting of two neutrons and one proton
H is an ordinary hydrogen nucleus.

Both of these reactions are approximately equally probable. Unfortunately, extremely high temperatures, perhaps 500,000,000°K, are required to make this fusion process self-sustaining (i.e., to continue without the addition of energy from the outside). Although there may be some fusion between a very few, extremely high velocity nuclei at temperatures below 500,000,000°K, the number of nuclei involved is not sufficient to maintain the temperature.

Because of the extremely high temperature required for the D+D process, most of the effort now being expended on fusion is being directed towards the deuterium-tritium process. It appears that this process will be self-sustaining at temperatures of about 100,000,000°K.

The fusion reaction of deutron and triton is as follows:

$$D + T \rightarrow He^4 + n + 176.\,\text{Mev} \qquad (19\text{-}3)$$

He^4 is the isotope of helium composed of two neutrons and two protons. Tritium does not appear naturally in water but may be obtained by bombarding lithium with neutrons. As stated later, the walls of the fusion chamber may be lined with lithium. Thus, during the fusion process, tritium is produced from the lithium walls. As with other minerals, it is not possible to evaluate our total supply of lithium. Dr. Samuel Glasstone states that ample reserves are available on land in the United States at a moderate cost, sufficient to last for hundreds of years and probable reserves would last for a few thousand years.† He states that even larger amounts are present in the oceans. Assuming that Dr. Glasstone is correct, the deuterium-tritium fusion process has the potential for supplying our energy needs for at least many centuries. As we gain knowledge from the operation of this process, it is possible that we will

* Mev (million electron volts) is defined in Section 9.1. It is suggested that the reader review Section 9.1 at this time.

† *Controlled Nuclear Fusion*, Samuel Glasstone, United States Atomic Energy Commission, 1974.

ultimately be able to develop the deuterium-deuterium process, thus giving us the potential for satisfying our energy needs for as long as we can contemplate. The problem now facing us is how we can bring the deuterium-tritium process into reality.

Although it has been demonstrated with the hydrogen bomb that we have the technology to carry out a fusion process, we have not demonstrated outside the laboratory that we can control such a process. In addition to heating the plasma to the desired temperature, it must be confined for a sufficient time for the fusion process to take place. Assume that the plasma occupies part of a chamber whose walls are cooled. Unless prevented from doing so, the plasma will spread out and occupy the entire chamber. This will cause two adverse effects. As the plasma spreads out, there is less opportunity for the high velocity neutrons to come in contact with each other and hence there will be a decrease in the reaction rate. In addition, those neutrons which come in contact with the walls will be cooled to temperatures below that necessary for fusion to take place.

19.3 MAGNETIC CONFINEMENT

Consider a plasma existing in a cylindrical tube. Ordinarily, the particles constituting the plasma move along straight line paths in random directions with random velocities. If a uniform magnetic field is placed around the cylinder, this field will cause the particles to move in a cork-screw path. The positively charged particles spiral in one direction and the negatively charged ones in the opposite direction. Provided that the field is of sufficient strength, the charged particles do not strike the walls of the cylinder.

Even though the charged particles are kept from striking the walls by the magnetic field which surrounds them, heat will be lost to the walls by radiation. Thus, the fusion process will be self-sustaining only when the heat produced by fusion at least equals that lost by radiation. The heat produced by fusion is a function both of the temperature and the number of charged particles present per unit volume. Normally, the number of charged particles per cubic centimeter in the plasma now being considered ranges from 10^{14} to 10^{16}. This may be compared with approximately 3×10^{19} molecules of a gas at normal room conditions. Although the particle density (the number of particles per unit volume) for this plasma is much less than that for the molecules of a gas at room conditions, the pressure of the plasma is very much higher because of its extremely high temperature. For the normal plasma particle density, the pressure may be around 200 pounds per square inch. This is the pressure which the magnetic field must overcome to prevent the plasma coming in contact with the walls. Even though extremely strong magnetic fields are used for this purpose, it is not presently reasonable to try to confine plasmas having pressures greatly higher than about 200 pounds per square inch. This limiting pressure prevents the use of higher particle densities which would increase the rate of fusion.

It is possible to prevent the charged particles from leaving the cylinder at its ends by creating extra strong magnetic fields at the ends, thus reflecting the particles back into the cylinder. This process is called the mirror effect. Although the technical feasibility of the mirror effect has been demonstrated, most of the magnetic containment efforts have been directed toward placing the plasma in a doughnut-shaped container, known as a torus. The magnetic field causes the charged particles to move around and around within the torus. The behavior of a plasma in a strong magnetic field at extremely high temperatures is very complicated. There may be plasma diffusion as well as plasma instability. The magnetic field has a tendency to pinch the plasma stream. The nature of the pinch effect is variable, with the so-called low-beta and high-beta pinch effects being noted. The study of all of these effects is much too involved to be discussed here. It can be noted, however, that because of the behavior of the plasma in the strong magnetic field, various cross-sectional shapes of the torus have been devised which give better performance than does a circular cross-section. At this time, in both Russia and the United States, plasma temperatures in laboratories have been achieved which are one order of magnitude below that required for self-sustained fusion. It is felt by some that self-sustained fusion may be obtained by 1985.

Although with magnetic containment the plasma does not come in direct contact with the containing walls, the walls do receive much heat from the plasma by radiation. Furthermore as shown in Equations 19-1 and 19-3, neutrons are released as a result of the fusion process. Since the neutrons are not charged, the magnetic field does not prevent them from hitting the walls, bringing their energies with them. Various configurations may be used for the walls of the containing chamber (the reactor). One such arrangement, as discussed by Glasstone,* has the reactor chamber surrounded by a blanket of molten lithium or a lithium salt. The neutrons react with lithium to produce tritium, which is withdrawn and then fed to the reaction chamber. The temperature of the lithium may reach 2,000°F. The lithium blanket is contained in a blanket shell generally made of a refractory metal such as niobium, molybdenum or vanadium. This blanket is surrounded by a thermal insulator to absorb any stray neutrons which pass through the blanket shell in addition to providing the necessary insulation. Outside the thermal insulator is the magnetic containment coils.

Assuming that the fusion process can be made self-sustaining, the problem becomes one of choosing a method to use the heat generated. Glasstone suggested introducing a liquid metal, such as potassium, through pipes inbedded in the lithium blanket. The potassium will be vaporized and may be used to drive a potassium vapor turbine. Since the temperature of the vapor leaving this turbine is high, it may be used to produce steam for use in a steam turbine. Such an arrangement is shown in Figure 19-1. Since the temperature range for

* See footnote, Section 19-2.

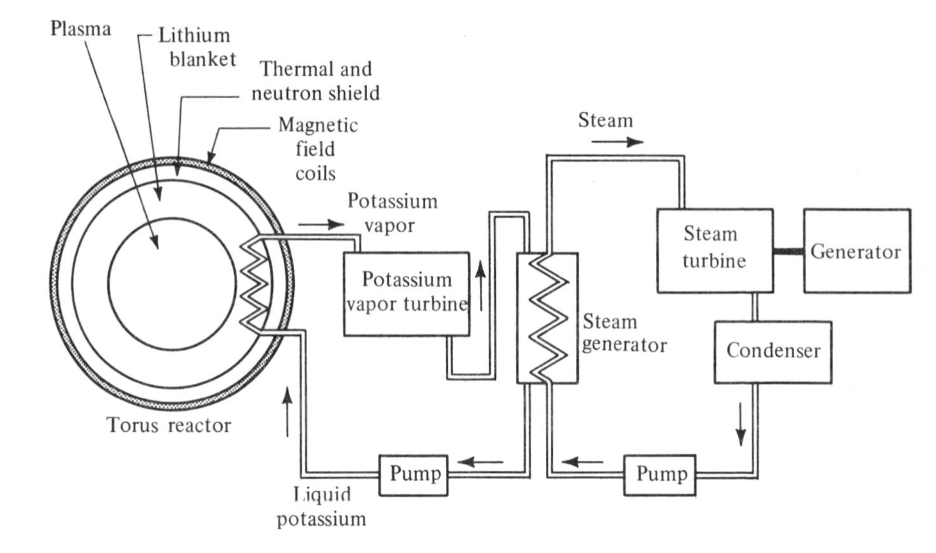

Figure 19-1. Elements of a suggested fusion power plant

the two turbines is high, it should be possible to attain thermal efficiencies of at least 50 percent. However, a considerable portion of the power developed normally must be used for the containment magnets. Because of this, consideration is being given to the use of superconducting materials for the magnets. Certain metals and alloys have their electrical resistance reduced to substantially zero when cooled below 10°K. Once current is induced to flow under these conditions, the flow continues even after the source of current is removed. Thus the power requirement for the confinement magnets can be reduced to a very low value. However, much energy must be expended to maintain the extremely low temperatures. The overall problem of maintaining a temperature in the reactor of around 100,000,000°K and a temperature close to absolute zero in the magnetic coils is only one of the major problems involved in fusion power production.

Several methods have been suggested for heating the plasma to the desired temperature. In one method, an electrical current is passed through the plasma. Because of its electrical resistance, the plasma becomes heated to temperatures as high as a few million degrees °K. At these high temperatures, the resistance of the plasma becomes so low that resistance heating cannot produce much further increase in temperature. The plasma temperature can be increased further by suddenly increasing the strength of the magnetic field. This produces magnetic compression, which increases the plasma temperature similar to that produced by ordinary compression of a gas. Consideration is also being given to the introduction of very high velocity neutrons into the plasma to heat it.

19.4 LASER FUSION

Another approach to achieving fusion is to use laser beams. (A laser is described as a device, usually containing a crystal such as a synthetic ruby, in which atoms, when stimulated by focused light waves, amplify and concentrate these waves, then emit them in a narrow, very intense beam.) In the laser fusion method (which is sometimes called inertial confinement), a pellet of deuterium-tritium mixture is injected into the reactor chamber. (The pellet is very small, being perhaps a millimeter in diameter.) A high energy laser beam, when focused on the pellet, causes an "explosion" and creates a plasma on the outer surface. The resulting "blow-off" causes the remainder of the pellet to implode. The converging shock compresses the remainder of the pellet and also heats it to the very high temperature required for a D-T reaction. This reaction generates so much pressure that the pellet explodes. Much of the energy produced is carried away by the neutrons formed during the fusion process. A new pellet can then be injected and the process repeated. The inertia of the pellet is depended on to provide the desired confinement. This means that the reaction must take place in a very short period of time—probably less than a nanosecond (a billionth part of a second). Thus a very high energy pulsed laser beam is required for the laser fusion process.

Such laser beams have not been developed as yet for actual operation but developmental work looks very promising. Some work is being done on using several laser beams, each focused on a given pellet, thus requiring much less energy per beam. Although it has been demonstrated in the laboratory that laser beams can be used to produce D-T fusion, the energy presently derived from the fusion process is less than the energy input to the lasers. Because of its potential, both government and industry support is increasing for the laser-fusion process.

19.5 FUSION SAFETY

Many of the safety hazards associated with nuclear fission power plants are absent from the fusion plants. There are no spent fuel elements to be transported and reprocessed, thus eliminating the possibility of theft of radioactive material. The fuel for the fusion reactor cannot be used for bomb construction. There are no long-life radioactive wastes to be disposed.

There are, however, some hazards inherent with a fusion reactor. An accident with the magnetic system could release as much energy as that of an average lightning bolt. The lithium blanket contains perhaps 300 times as much energy, which would be released in the case of a lithium fire. A much greater potential hazard is that of the tritium gas within the reactor. Tritium is radioactive and has a half life of 12 years. It is very difficult to contain. It has been estimated that at least 0.03 percent of the total inventory of tritium will escape per year. Much more tritium will penetrate the reactor structure. By proper design, this tritium should not escape to the outside but will remain in the reactor structure if it is disassembled. Some tritium is expected to be found in the water released in the steam cycle.

Of course, the magnitude of the harmful effects which can be caused by tritium cannot be judged until full-size fusion reactors are built. It appears that the fusion reactor power plant will present many fewer hazards than the fission power plants. Nevertheless, there are sufficient hazards associated with fusion power plants that make their location in built-up areas questionable.

19.6 SUMMARY ON NUCLEAR FUSION

It is recognized that nuclear fusion offers the potential for supplying our energy needs almost indefinitely, provided that the deuterium-deuterium process can be used. Some problems involved in fusion can be reduced a little if the deuterium-tritium process is used. At present, tritium is obtained from lithium, the reserves of which are large but not unlimited. Hence the potential for power production using the deuterium-tritium process is somewhat limited. However, tremendous problems—almost beyond comprehension—must be solved before fusion power can become a reality. In spite of growing resources being devoted toward the development of both the magnetic containment and the laser fusion process, not only here, but in foreign countries, particularly Russia, most authorities feel that fusion power cannot become a commercial reality before the year 2000. There are a few skeptics who wonder if fusion power will ever become practical.

Although there are some hazards associated with nuclear fusion, these hazards appear to be small when compared with those connected with nuclear fission. Because of the great potential of fusion power, it is most desirable that we devote increasing amounts of our resources to its development.

SUPPLEMENT

PROBLEMS

19-1. Assuming that the deuterium-deuterium fusion reaction can be fully developed and the heat produced utilized for power production with a thermal efficiency of 50 percent, determine the number of cubic feet of water required per year to produce our anticipated electrical energy needs for 1990. (See Table 2-3). Make use of the relationship between the amount of water and gasoline for a given energy content as given in Section 19.2. Take the heating value of gasoline as 145,000 Btu per gallon. Also express the volume of water required per year in cubic miles.

19-2. Do you feel that the government should devote billions of dollars per year to the development of fusion power? Why?

19-3. Do you feel it reasonable to expect that private industry will, in the future, invest large sums of money in the development of fusion power? Why?

19-4. Discuss the desirability of using our resources to develop the fast breeder fission reactor rather than to develop fusion power.

19-5. Discuss the desirability of concentrating our resources on the development of fusion power rather than on solar energy, wind energy, geothermal energy and thermal gradients of the oceans.

19-6. Since deuterium is relatively inexpensive, why is it desirable to install an efficient but expensive power producing system, such as shown in Figure 19-1?

19-7. Since it has not been demonstrated even in the laboratory that fusion power is possible, should the federal government stop funding all work infusion power?

19-8. In consideration to the answer in Problem 19-1, do you feel that the world will never have to worry about energy sources if nuclear fusion becomes economically feasible?

PART IV

ENERGY TRANSFORMATIONS

20

Coal Gasification and Liquefaction

20.1 INTRODUCTION

It has been stated earlier that oil and natural gas supply approximately 72 percent of our present energy needs. But these fuels domestically are in short supply and may be unable to make a very large contribution toward meeting our energy needs two or three decades from now. On the other hand, we have enormous reserves of coal, probably sufficient to last at least several centuries. Unfortunately, it is difficult to use coal directly to supplant oil and natural gas in most of their usages. It is possible, however, to gasify or liquefy coal, thus offering the potential of substituting, indirectly, our very large coal resources for the much smaller reserves of natural gas and oil.

The concept of coal gasification is an old one. A so-called gas producer was in operation as early as 1836. In the mid 1920's there were approximately 21,000 gas producers in operation in this country, using some 25 million tons of coal per year. This amount of coal is almost 4 percent of all the coal now used annually in this country for all purposes. Gas producers, however, were relatively bulky, inefficient and produced a low Btu gas, although it was frequently enriched by the addition of oil which was carbureted. With the development of large natural gas fields, particularly in Texas and Louisiana, and with the spread of natural gas pipelines throughout the country, low cost, high Btu content gas became available almost everywhere. Not being able to compete with natural gas, gas producers gradually were phased out.

In recent years, with the increase in demand for gaseous fuels and with the growing shortage of natural gas, an increasing effort is being devoted to coal gasification.

The concept of coal liquefaction also is not new. During World War II, the Germans, with very little petroleum available to them, developed a process for producing gasoline from coal.

20.2 GASIFICATION CONCEPT

As stated in Chapter 6, coal consists of moisture, volatile combustible, fixed carbon, and ash. The relative percentages of these constituents vary widely, depending on the type of coal. The combustible portion of anthracite coal is largely fixed carbon. On the other hand, only about half of the combustible portion of lignite is fixed carbon. If a lignite coal is heated sufficiently in the absence of air, about half of its combustible portion will be driven off, thus producing a gaseous fuel. However, the simple heating of an anthracite coal will produce very little gas. A chemical action, such as combustion, is required to convert the fixed carbon in a coal into a gas. It is evident then, that the nature of the gasification process to be used is dependent not only on the proposed use of the gas produced but also on the nature of the coal to be used.

It is the purpose of this chapter to describe the general concepts of coal gasification rather than to discuss the chemistry of each reaction and details of the equipment used in the gasification process.

The various types of gas produced from coal can be classified into three groups: (1) low Btu gas, (2) intermediate Btu gas, and (3) high Btu gas. As will be discussed, low Btu gas shows its greatest potential for use in power generation. For this reason, it is frequently known as *power gas*. Intermediate Btu gas is suitable for industrial use and hence is known as *industrial gas*. Only high Btu gas is suitable for transmission through long pipelines. Frequently it is known as *pipeline gas*.

20.3 POWER OR LOW BTU GAS

When a coal is burned with a deficiency of air, some carbon monoxide is produced. This is a simple but inefficient way to produce a combustible gas from coal. Furthermore, the heating value of this gas is very low. A more realistic method is to supply steam to the coal as it is burned with a deficiency of air. This process is commonly known as the steam-air blown process. During this process, the steam is dissociated into hydrogen, which becomes one of the constitutents of the gas delivered from the process, and into oxygen, which combines with the carbon in the coal to form carbon monoxide and some carbon dioxide. Chemical actions involving the volative matter in the coal will also take place. The final composition of the gas produced is dependent on the nature of the gasifier, on the amount of air and steam supplied and on composition of the coal. A typical analysis of the gas shows 4 to 5 percent carbon dioxide, 25 to 27 percent carbon monoxide, 12 to 13 percent hydrogen, 3 to 4 percent methane and similar hydrocarbons, and 50 percent or more of nitrogen. Because of the nitrogen and carbon dioxide present, the heating value of this gas is low—perhaps 100 to 200 Btu per cubic foot, compared with over 1,000 Btu per cubic foot which is normal for natural gas.

Although some of today's gasifiers use the fixed-bed concept,* which was used by the old gas producers, the use of fluidized beds are quite common. In this method the bed is agitated by air forced through it at a sufficiently high velocity to make the bed turbulent and thus increases the opportunity for the reaction to take place more rapidly. In other gasifiers, the coal particles are held in suspension as the chemical reaction takes place.

Because of its low heating value it is not economical to transmit this gas more than a very few miles from the point where it is produced. Thus it shows little potential for supplanting natural gas for residential and commercial heating purposes and for light industrial use when the users are located at some distance from the gasifier. There are, however, three potential areas for use of power gas.

A potential use of power gas is for steam power plants. As stated in Chapter 6, much of the coal burned in this country is burned in the east. Unfortunately, most of the eastern coal contains excessive amounts of sulfur. Although methods are available to reduce the sulfur content of the resultant gases produced by combustion to acceptable values, the present cost of doing so is high. The alternative is to gasify the coal at the power plant site. Since the volume of the gaseous fuel is much smaller than that of the gases produced by combustion, it is easier and cheaper to remove the sulfur from the gaseous fuel. Since power-gas gasifiers of the size required for large power plants have not been built and since methods for removing sulfur either from the coal as it is burned or from its products of combustion have not been fully developed, the potential for the use of power gas for steam power plants cannot be completely evaluated. It should be noted however, that the cost of the gasifier will be high. Furthermore, some gasifiers have a relatively low efficiency, perhaps around 70 percent (i.e., the energy of the gas delivered being around 70 percent of that of the coal). Much higher efficiencies have been reported for certain small pilot plant operations. It is questionable whether these high efficiencies can be attained in large scale plants. By integrating the operations of the gasifier with that of the power plant, it may be possible to recover some of the energy lost in the gasification process.

A second potential use of power gas is for gas turbines. Gas turbines in power plants do produce significant amounts of power, particularly for peak load conditions. A considerable effort is being expended to design gas turbines to operate at higher temperatures, thus at increased efficiencies. It is expected that gas turbines will be used more extensively for power production, especially when used in dual cycle operation. (See Chapter 22.) In the past, natural gas was widely used as a gas turbine fuel. Because of growing shortages, the use of natural gas for this purpose is being greatly curtailed. Today, various grades of fuel oil supply most of the energy used by power plant gas turbines.

* In the fixed-bed method, coal is fed onto grates. Reactions take place as steam and air are supplied through the grates.

But oil is also in short supply and becoming evermore expensive. Hence, for large installations, it seems possible that power gas gasifiers may become economically feasible for producing fuel for power plant gas turbines.

When a group of industries, particularly large ones, are located close together, it may be feasible to install a low Btu or power gas gasifier. Certain industries, such as some textile operations, fertilizer producers, and brick makers, have been designed to operate on a gaseous fuel. Formerly, with natural gas cheap and abundant, many of these industries were able to obtain natural gas at a very low price by signing an interruptible contract. Such a contract means that these industries have the lowest priority for the available gas and are now facing shut-downs in the winter months because of the shortage of natural gas.

20.4 INDUSTRIAL OR INTERMEDIATE BTU GAS

The prime reason for the low heating value of power gas is that air is used for its production. The nitrogen present in the air appears in the resultant gas, thus greatly diluting it. A gas having a much higher heating value can be produced by supplying the gasifier with oxygen rather than air. Such a gasifier may be termed a steam-oxygen blown gasifier. The heating value of the gas thus produced will be approximately forty percent of that of natural gas.

Because of its larger heating value, it is economically feasible to transmit it over somewhat longer distances than is the case for power gas. When several smaller industries, requiring sizeable amounts of gas, are located within a few miles of each other, it may be feasible to install a centrally located industrial gasifier. This may become more attractive when natural gas becomes more expensive and less available to industries during winter months.

It should be noted that, on a Btu basis, the cost of industrial gas will be higher than that of power gas since an oxygen producer must be installed and operated to produce the oxygen required to make the industrial gas.

20.5 PIPELINE OR HIGH BTU GAS

Much of our natural gas is used for residential and commercial heating. As such, the amount used by a given customer is relatively small. This means that the gas must be piped relatively long distances—much too long to consider using industrial gas as an alternative.

Natural gas is largely methane, CH_4. To produce a gas rich in methane, coal, oxygen and steam are supplied to a gasifier, similar to the gasifier for the production of industrial gas. The resultant gas generally is cleaned, with the sulfur and perhaps the carbon dioxide being removed. It is then fed into the methanator. Here, by catalytic* action, hydrogen is combined with carbon to form methane. High temperatures and high pressures are required for satisfac-

* A catalyst is a substance which causes or speeds up a chemical action without itself undergoing any permanent change.

tory operation. In certain cases, temperatures as high as 2,000°F to 2,700°F and pressures as high as 1,000 psi are used.

The hydrogen required for the methanation process may be supplied from an external source but it is normally obtained by the decomposition of steam. This may be accomplished by reacting the steam with carbon monoxide. This reaction produces carbon dioxide and free hydrogen, which can be fed into the methanator.

The final gaseous product leaving the methanator is largely methane. As such it is similar to natural gas and can be substituted for it. Just as with natural gas, it is economical to transmit it long distances. If pipeline gas becomes a commercial reality it can be readily introduced into the vast network of existing natural gas pipelines.

The overall efficiency of a pipeline gasifier is low, ranging from about 55 to 70 percent. The initial cost of the gasifier is much higher than other types of gasifiers since the methanator is added equipment and must be designed to operate at high temperatures and pressures. Hence, the cost of pipeline gas is expected to be much higher than other types of gas produced from coal and very much higher than the present cost of natural gas and possibly even higher than the cost of natural gas a decade from now. Although cost estimates have been made for gas produced in various types of gasifiers, meaningful figures will not become available until full-size units are operated. Regardless of costs, however, when natural gas becomes unavailable for certain usages, we may have to use manufactured gas for those purposes where a gaseous fuel is most satisfactory.

Before leaving the subject of coal gasification, one fact should be mentioned. Much of our eastern coal is a caking coal (i.e., upon heating, the coal fuses together into solid cakes). Unless means are provided to break up the cakes, this type of coal cannot be handled satisfactorily in some types of gasifiers.

20.6 UNDERGROUND (IN-SITU) COAL GASIFICATION

Underground mining of coal is expensive and, unless many safety precautions are made, also dangerous. To avoid these problems, gasifying the coal underground has been proposed. Here, the general concept is to drill holes into the coal bed through which air is supplied. Another set of holes, the gas holes, are drilled at some distance from the air holes. The gases resulting from combustion seep through the coal bed to the gas holes, from which they are removed. To permit movement of the gases through the coal bed, cracks must exist in the coal or must be created by some type of explosion.

Underground gasification has been carried out on a small scale for many years both in this country and in foreign countries, thus demonstrating its technical feasibility. There are, however, many serious problems to be solved before this method of gasification can become a commercial reality. One of the major problems is that of controlling the combustion so that a gas of the desired quality can be produced. Involved in this problem is the tendency for burn-through* to occur near the center of the bed, making it difficult to burn the remainder of the coal in the bed.

Because of the potential of this method of coal gasification, investigational work is being continued in an effort to solve some of the major problems.

20.7 OIL GASIFICATION

Some effort is being devoted to the gasification of petroleum products, particularly naphtha. The technology for carrying out this gasification process exists and is being used to a limited extent to produce a high Btu gas, which can be substituted for natural gas. Although it is anticipated that some gas will be produced from oil in the next decade or two, because of the growing shortages of oil, it does not seem reasonable that much gas will be produced in this way after that time. For this reason, the details of the oil gasification process will not be considered here.

20.8 COAL LIQUEFACTION

The carbon content of coal weighs approximately 16 times more than that of the hydrogen. In petroleum products, the carbon-hydrogen ratio is about six to one. Thus, to liquefy coal, much hydrogen must be added to the carbon to form large molecules, generally of the form C_nH_{2n}, if a liquid fuel is to be produced. As with coal gasification, there are various methods to liquefy coal. In general, these methods can be classified as follows:

1. Hydrogenation
2. Pyrolysis
3. Catalytic Conversion

Hydrogenation of coal can be accomplished by adding either hydrogen as a gas or a hydrogen-rich liquid solvent. Particularly when a solvent is used, a catalyst is required and the reaction must be carried out at temperatures up to 900°F and pressures up to 4,000 psia. When gaseous hydrogen is used, the products may include, in addition to a liquid fuel, a gaseous fuel and also char.

When the amount of volatile combustible in the coal is high, the pyrolysis method shows much promise. In this process, the coal is heated in the absence of air, thus driving off the volatile matter. The volatile matter may be cleaned, the sulfur removed, and then hydrogenated to produce a liquid. The char may be burned either to produce power or to furnish the heat required for the pyrolysis process. As an alternative, the char can be burned with a controlled amount of oxygen and steam to produce a gas somewhat similar to an industrial gas.

In the third method, an industrial gas is first produced from the coal. The hydrogen and the carbon monoxide in the gas are then combined, after clean-

---

* A hole may be burned through the bed, thus permitting the air supplied for combustion to flow straight through the bed rather than to burn additional coal.

ing, by the use of a catalyst to form a liquid fuel. It is essential that the proper ratio of hydrogen to carbon monoxide is present to permit the liquefaction to take place.

There are two broad areas for use of liquid fuels produced from coal. One of these is for power production, particularly in steam power plants. Because of the high sulfur content of a very large percentage of our eastern coals, there has been a shift from coal to oil and, where available, to natural gas for power generation. In fact, some localities have prohibited the burning of coal. The growing shortages and rapidly increasing prices of both oil and natural gas make it desirable to consider the liquefaction of coal to produce a fuel for this purpose. Since the power plants do not require a highly refined product of restricted volatility and viscosity, the liquefaction plant to produce a fuel for power plant use should be relatively simple and the fuel produced relatively inexpensive.

Most of our petroleum products are used in transportation and in space heating, particularly residential heating. Gasoline, diesel fuel oils and home heating oils are highly refined and have rather restrictive specifications. Thus a coal liquefaction plant producing an oil which can be refined for these uses must be more elaborate and hence more expensive to build and to operate.

20.9 SOLVENT REFINING

Another method of producing a "clean" fuel from coal is that of solvent refining. There are several different processes used in solvent refining. In general, pulverized coal is dissolved in a coal-derived solvent (such as phenanthrene) in the presence of hydrogen at high temperatures and pressures (perhaps 800°F, 1,000 psia). After residue separation (i.e., the removal of the inert materials which do not dissolve) the material left is known as the filtrate. The filtrate is now heated to recover the solvent and some oil. The major portion of the filtrate does not evaporate but remains largely as a solid although some thick liquid may be present, depending on operating conditions. This solid is known as solvent refined coal. It contains very little sulfur and very little ash and has a heating value, on the pound basis, materially higher than that of the coal from which it was obtained. It is thus very satisfactory as a boiler fuel.

If the desired end product is primarily oil, the solvent refined coal may be hydrogenated at high pressures and temperatures in the presence of a catalyst in much the same way as was described for liquefying coal. The oil thus produced from the refined coal will not contain any appreciable amount of sulfur.

20.10 THE STATUS OF COAL GASIFICATION AND LIQUEFACTION

Many problems remain to be solved before coal gasification and liquefaction can make a significant contribution to our energy supply. Some of these problems are

1. The production and transportation of the coal
2. Losses of energy during the gasification and liquefaction processes
3. High initial costs of the plants
4. High operating costs
5. Environmental considerations

The problems of mining and transporting the coal were discussed in Chapter 6. These problems involve environmental considerations, mine safety, financing of the expanding mining operations, skilled labor required for the mining operations, and equipment and manpower required to transport the coal.

As stated earlier, the efficiency of gasifiers, particularly the pipeline gas gasifiers, is low. Much research and development is required to improve the efficiency, and thus reduce the cost of the fuel delivered.

A considerable amount of work has been done towards predicting both the initial and operating costs of full-size plants. These projected costs, particularly for the pipeline gas plants, are high—too high to be competitive with other fuels in the near future. Without a doubt the first full-size plants will be very expensive since they will involve development costs. But studies must be made concerning how the intital costs of future plants and their cost of operation can be reduced.

In addition to the environmental problems associated with the production of the coal, there are several environmental problems associated with gasifiers and liquefiers. Much water will be required for their operation. This may be a real problem in the drier parts of the country. Unless proper precautions are taken, the waste water will carry contaminants away with it. Finally, the solid wastes from the plants must be disposed of with care since they may cause pollution.

Several industrial companies here and abroad as well as the U.S. Bureau of Mines have been working for some time on various types of pilot plants for gasifying and liquefying coal and on solvent extraction. The technical feasibility of all of these processes has been demonstrated. Some commercial plants are now being planned.

Without factual information from full-size plants it is very difficult to predict accurately the contributions which will be made to our energy supply by coal gasification and liquefaction. Many predictions however, have been made. One such prediction is given in "Synopsis of Energy Facts and Projections.* In this report three sets of predictions are made, namely High Trend, Medium Trend and Low Trend. The Medium Trend prediction for coal gasification and liquefaction is given in Table 20-1.

According to this prediction, the energy in the coal used for gasification and liquefaction in 1995 will be approximately two percent of the total energy used in this country. However, vast sums of money must be raised to expand our coal mining facilities and our coal transportation facilities. Enormous sums of

---

* U.S. Department of Energy, "Synopsis of Energy Facts and Projections," *1978 Annual Report to Congress*, p. 47, Washington, D.C.

Table 20-1. Predicted Coal Gasification and Liquefaction (Millions of Tons per year)

	1985	1990	1995
Total, all purposes	960	1,385	1,911
For gasification	7	35	74
For liquefaction	6	5	56

money and much of our resources also will be needed to construct gasifier and liquefier plants.

20.11 SUMMARY ON COAL GASIFICATION AND LIQUEFACTION

At present oil and natural gas satisfy approximately 72 percent of our energy needs. Our demands for these excellent fuels are ever-increasing. But our production of these fuels is decreasing and, it appears, our reserves will be greatly depleted in two or three decades. To meet our demands for these fuels, particularly oil, we are importing vast quantities, putting a strain on our economy. This situation probably will become much worse. On the other hand, we have large sources of coal. We have developed the technology to both gasify coal and to liquefy it. Various types of pilot plants have been constructed and plans are being developed for some commercial plants. Many problems exist, however. Present conversion efficiencies are relatively low. Very large sums of money will be required to construct these conversion plants. We must greatly expand our coal mining and coal transportation facilities. Major environmental problems must be solved. It appears that sufficient steps can be taken in the next two or three decades so that significant amounts of gas and oil will be produced from coal, thus helping to alleviate the shortages of these fuels. However, the costs of these substitute fuels will be high, probably, at least at first, higher than petroleum and natural gas.

SUPPLEMENT A

ILLUSTRATIVE EXAMPLES

Example 20-1. A power plant, having an overall efficiency of 38 percent, delivers 1,200,000 kW. It uses coal having a heating value of 13,200 Btu per pound. Since the coal is high in sulfur, it is first gasified to remove the sulfur. The efficiency of the gasifier is 73 percent. Determine the number of tons of coal required per hour.

Solution. The output is equivalent to $1{,}200{,}000 \times 3{,}413$ Btu per hour. The output equals $13{,}200 \times 0.38$ Btu per pound of coal or $13{,}200 \times 0.38 \times 2{,}000$ Btu per ton of coal.

If the coal were to be burned directly the tons per hours, then, equals

$$\frac{1{,}200{,}000 \times 3{,}413}{13{,}200 \times 0.38 \times 200} = 408.3 \text{ tons}$$

Since the efficiency of the gasifier is 73 percent, the amount of coal required for it equals

$$\frac{408.3}{0.73} = 559.3 \text{ tons per hour} \qquad \text{answer}$$

Example 20-2. A pipeline gasifier is to produce 2,500,000 cubic feet of gas per hour, having a heating value of 1,010 Btu per cubic foot. It has an efficiency of 68 percent. It is supplied with coal having a heating value of 12,800 Btu per pound. Determine the number of tons of coal required per hour.

Solution. The heating value of the gas produced per ton of coal equals

$$12{,}800 \times 0.68 \times 2{,}000$$

The heating value required equals

$$2{,}500{,}000 \times 1{,}010$$

Hence the tons of coal required per hour equals

$$\frac{2{,}500{,}000 \times 1{,}010}{12{,}800 \times 0.68 \times 2{,}000} = 145.0 \text{ tons} \qquad \text{answer}$$

Example 20-3. The coal in Example 20-1 has a sulfur content of 3.4 percent. If 80 percent of the sulfur is removed during the gasification process, determine the reduction in the amount of sulfur which will be discharged into the atmosphere per week by gasification rather than direct burning.

Solution. Sulfur discharged per hour, direct firing, equals

$$408.3 \times 0.034 = 13.88 \text{ tons/hr.}$$

Sulfur discharged per hour, gasification, equals

$$559.3 \times 0.034(1 - 0.8) = 3.80 \text{ tons/hr.}$$

Reduction in sulfur discharged per week equals

$$(13.88 - 3.80)24 \times 7 = 1{,}693 \text{ tons/week} \qquad \text{answer}$$

Example 20-4. A coal gasifier produces industrial gas having a heating value of 430 Btu per cubic foot. The gasifier efficiency is 74 percent. Determine the volume of gas produced per hour when burning 125 tons of coal per hour. The heating value of the coal is 12,720 Btu per pound.

Solution. The cubic feet of gas produced per pound of coal

$$\frac{12{,}720 \times 0.74}{430} = 21.89 \text{ cu. ft.}$$

The gas produced from 125 tons equals

$$21.89 \times 125 \times 2{,}000 = 5{,}473{,}000 \text{ cu. ft./hr.} \qquad \text{answer}$$

Example 20-5. A pipeline for pipeline gas has a diameter of 36 inches. Determine the diameter required to handle power-gas having the same total energy content. Assume equal gas velocities. The heating value of pipeline gas is 1,010 Btu per cubic foot and the heating value of the power-gas is 180 Btu per cubic foot.

Solution. The pipeline areas vary inversely with the heating value. The ratio of the area of the power gas line to that of the pipe for the pipeline gas equals

$$\frac{1{,}010}{180} = 5.61$$

The diameters vary as the square root of the areas. Thus the diameter power-gas pipeline equals

$$36\sqrt{5.61} = 85.3 \text{ inches} \qquad \text{answer}$$

SUPPLEMENT B

PROBLEMS

20-1. a. It is proposed to substitute power gas for coal in a power plant burning 450 tons of coal per hour. The efficiency of the gasifier is 72 percent. Determine the number of extra tons of coal to be supplied per day because of the losses in the gasifier.
b. The coal in part a costs 40 dollars per ton. Calculate the yearly cost of the extra coal used because of the gasifier.

20-2. The coal in Problem 20-1 is 10.2 percent ash. Calculate the extra tons of ash to be disposed of because of gasification.

20-3. A coal having a heating value of 13,200 Btu per pound is used to produce pipeline gas having a heating value of 1,000 Btu per cubic foot. The efficiency of the gasifier is 65 percent. Determine the number of cubic feet of gas which can be produced per ton of coal.

20-4. Considering the answers obtained in Problems 20-1, 20-2, and 20-3, do you feel that when only high sulfur coal is available, the coal should be gasified for power plant use rather than burned directly? Why?

20-5. a. Low sulfur coal is available. Should this coal be gasified for use in steam power plants? Why?

b. Same as part a but for gas turbine power plants.

20-6. How many tons of coal must be liquefied per year to provide 10 percent of the energy predicted to be supplied by oil in 1985 (see Table 2-1). Assume the heating value of coal to be 12,800 Btu per lb. and the conversion efficiency to be 74 percent.

20-7. Same as Problem 20.5 but for a high sulfur coal.

21

The Hydrogen Economy

21.1 INTRODUCTION

There are some indications that, in the next two or three decades, hydrogen may replace oil and natural gas in many areas of our society, including residential and commercial heating, transportation, industrial uses, and even, perhaps in electric power generation. The term "hydrogen economy" has been introduced to designate this use of hydrogen.

In many ways, hydrogen is an ideal fuel. It is much easier to burn than solid and liquid fuels. The product of its combustion is water vapor. When burned in air, some nitrogen oxides are produced from the nitrogen present in the air. The amount of these oxides is materially less than those normally produced when fossil fuels are burned. Tests carried out in single cylinder laboratory engines using gasoline showed the amount of nitrogen oxides produced to be several times that produced when the fuel was hydrogen.

Hydrogen is non-toxic but it is highly explosive. Many of those who recall the destruction by explosion of the hydrogen-filled airship Hindenburg fear the use of hydrogen as a fuel. Natural gas and gasoline, propane and butane vapors are also explosive and have caused some disastrous explosions. However, the chance of an explosion of any of these fuels is so very small that we do not hesitate to use them. In general, the chances for hydrogen explosions are lower than those of other fuels. To be sure, because of its small molecular size, it is difficult to contain. However, once it enters the atmosphere, it has a tendency to disperse rapidly. Since its density is so much lower than air, it cannot collect in low-lying pockets as do fuel vapors, such as those of gasoline. It should be noted that we presently use large quantities of hydrogen for various industrial processes. There have not been any significant number of accidents associated with the handling of this hydrogen.

Although there is some feeling among the general public that hydrogen can make a significant contribution towards solving our energy problems, hydrogen can make no direct contribution. It is not a primary fuel since it does not exist in the free form in nature. It is obtained only by expending energy obtained from other sources. Because of the inherent inefficiencies in the production of hydrogen, the energy of the hydrogen is significantly less than the energy expended in its production. Its potential value as a fuel lies in its ability to provide a means of transporting energy. This will be discussed in subsequent sections.

21.2 HYDROGEN PRODUCTION

Hydrogen can be produced from petroleum products and natural gas, both of which are rich in hydrogen. Approximately half of the hydrogen which is now used is obtained by this means. However, with the growing shortages of these fuels, they cannot supply the vast amounts of hydrogen anticipated for future use. Hydrogen can be obtained rather readily by electrolysis of water. In this method, a large number of electrolytic cells of various designs are used. These cells normally operate at a low voltage, perhaps one or two volts. Hydrogen is produced at the cathode (negative electrode) of these cells and oxygen at the anode (positive electrode). Although some electrolyzers operate at efficiencies around 50 percent, other commercial electrolyzers are reported to have efficiencies ranging from 60 to 70 percent. Efficiencies as high as 85 percent have been claimed in laboratory operations. It is felt by some that this efficiency can be ultimately attained in commercial operations.

If very large amounts of hydrogen are to be produced by electrolysis of water, so much oxygen will be produced that it can be sold at a cost materially less than that at present. A cheaper oxygen has the potential for many uses, such as for the production of industrial gas, pipeline gas, liquid fuel, magnetohydrodynamics (See Chapter 22), fuel cells, and polluted water treatment, as well as expansion of its present usages.

A second method of production of hydrogen is the thermal-chemical method. When water vapor is heated to extremely high temperatures (say above 5,000°R), a significant portion of it will break up (dissociate) into hydrogen and oxygen. However, the much higher temperatures required to dissociate most of the water vapor are too high to be practical. Work is being done on the thermal-chemical method of dissociation of water as an alternative to the direct thermal production of hydrogen. Various chemicals, such as calcium bromide, $CaBr_2$ are reacted with water, and the products which are produced then reacted with other chemcials to ultimately produce hydrogen and water, with the chemicals involved being reprocessed into their original form. The energy required for the dissociation of the water comes from the heat which must be added to cause the various reactions to take place. Efficiencies as high as 50 percent have been claimed for this method. Such an efficiency is, of course, much higher than the combined efficiency of turning heat into electrical energy and then using electrical energy to produce hydrogen by electrolysis. Since the thermal-chemical method has been demonstrated only in the laboratory, it is anticipated that many problems will be encountered and must be solved before this method can become a commercial reality. However, because of its reported higher efficiency this method has considerable potential.

Other methods have been suggested for the production of hydrogen, such as directly by the use of solar energy* or by algae reactions during photosynthesis of water. The feasibility of most of these methods has not yet been demonstrated even in the laboratory.

21.3 HYDROGEN USE FOR ENERGY STORAGE

As has been stated previously, it appears that electrical energy will produce an ever-growing percentage of our energy requirements. Thus it becomes more important than ever to attempt to generate electric energy at the lowest possible costs. The demand on conventional power plants varies greatly, with the nighttime demand being only a fraction of that during the daytime. But the fixed charges (See Chapter 4) go on at night at the same rate as in the daytime. If it were possible to operate the power plants at night at near full load and to store the excess energy for daytime use, the size of the plants to meet the daytime needs could be held down. With smaller power plants, the initial and hence the fixed costs would be reduced. Power, then, could be sold at a lower price.

Fossil fuels and nuclear fuels are relatively easy to store. The potential energy of water for hydropower production can be stored in reservoirs. But it is necessary to transform electrical energy to another form in order to store it. Although electric storage batteries can store small amounts of electrical energy, present batteries are not practical for the storage of large amounts. In certain regions it is practical to use pumped storage for this purpose, (See Chapter 10) but the regions where pumped storage can be used are limited. Under these conditions, much consideration is being given to the production of hydrogen by electrolysis of water during periods of low power demands. The hydrogen, then, could be used as needed.

As discussed in Chapter 12, there is a growing effort being made to develop solar energy for electrical energy production, either by means of solar cells or by steam power plants. It is expected that there will be considerable variation in the output of these plants, since there is much variation in the solar energy which is received throughout the day, and, of coure, there is no output at night. Thus, the potential exists for producing hydrogen by electrolysis at times of maximum electrical production and storing it for later use.

A somewhat similar situation exists with wind power (See Chapter 13). In many regions wind speeds are quite variable, creating the problem of energy storage for use during times of low wind speeds. Here again, the potential exists for producing hydrogen by electrolysis.

---

* Laboratory studies indicate the possibility of using solar energy to produce hydrogen directly from water by using a complex ruthenium compound as a catalyst.

21.4 HYDROGEN FOR ENERGY TRANSPORTATION

The trend in power plant design is to use increasingly large units and using more units in a given plant. This means that electrical energy will be transmitted over longer distances. Line losses, in these cases, become sizeable and the cost of lines contribute significantly to the final cost of electricity. Environmental objections are being raised as to the appearance of the transmission systems and their encroachment on land areas. There is an attempt to meet these objections by placing transmission systems underground. But, at present, the cost of an underground system is materially higher than the above ground one. Studies indicate that it will be more economical to produce hydrogen at the power stations and pipe it to the point of use for the longer distances—say 100 to 200 miles or more. This is particularly true when the hydrogen is to be used for heating purposes. The economics of such a use of hydrogen cannot be verified until actual installations are made.

Chapter 13 discussed the possibility of locating wind power plants off the east coast at a considerable distance. To avoid the difficulties of transmitting large amounts of electrical energy for a long distance beneath the ocean, it has been proposed to produce hydrogen at the wind power stations and to pipe it beneath the ocean to land. A similar proposal has been made for the method of using the energy produced by ocean thermal gradient power plants. (See Chapter 15.)

There is some feeling that hydrogen can readily supplant natural gas as gas supplies dwindle, with the hydrogen being fed into existing natural gas pipelines. However, trouble will be encountered if hydrogen is introduced into these pipelines. Hydrogen, having a density of only one-eighth that of natural gas, is much more difficult to contain. Leaks will occur in the pipeline and in the pumping equipment where none occurred with natural gas. In addition, hydrogen embrittlement may occur when present pipe steel is exposed to hydrogen under pressure.

The heating value of hydrogen on the basis of weight is much higher than that of other fuels. For example, the heating value of hydrogen is approximately 61,500 Btu per pound compared with approximately 20,500 Btu per pound for gasoline. On the other hand, because of its very low density, the heating value of gaseous hydrogen on the cubic foot basis of volume is only about one-third that of natural gas. This means that three times as many cubic feet of hydrogen gas must be transmitted to deliver the same energy that is deliverable by natural gas. Other things being equal, this would mean that much larger pipelines would be required for hydrogen. However, because of its low density and its low viscosity, the velocity of the hydrogen can be made much higher than that used for natural gas, thus permitting the use of pipelines for hydrogen of approximately the same size as those for natural gas. Because of its much larger volume, the pipeline compressors for hydrogen must be designed to handle these larger volumes.

21.5 USING THE ENERGY OF HYDROGEN

Hydrogen may be burned to produce heat or it may be used for power production. This is in addition to its many present uses. With the growing shortages of natural gas, hydrogen has the potential to substitute for natural gas for heating purposes in most applications. However, particularly for residential uses, a new network of pipelines would be required. Present natural gas burners would have to be modified if hydrogen is to be sutstituted. Because of the many factors involved, it is not possible, until it has been used on a large-scale basis, to evaluate the feasibility of using hydrogen as a replacement for natural gas for heating rather than to use pipeline gas made from coal.

The hydrogen-oxygen fuel cell seems to be a natural for using hydrogen for power generation (See section on Fuel cells in Chapter 22.) The technical feasibility of this cell has been demonstrated, particularly for auxiliary power production for spacecraft. There are indications that its efficiency can be greatly increased, perhaps to as high as 60 percent. As the hydrogen-oxygen fuel cell becomes mass produced, its costs may be reduced sufficiently to make it competitive with other power producing devices. Whether or not the hydrogen-oxygen fuel cell will be used extensively in the future will depend not only on its efficiency and cost but also on the source of hydrogen and its costs. Also to be considered are possible savings which can be made by locating fuel cells in regions of low power demands rather than to supply these regions by use of long distance power transmission lines, with their inherent line losses and high capital costs.

It is possible to burn hydrogen in the furnaces of steam power plants in place of fossil fuels. However, adjustments must be made in the furnaces to satisfactorily burn the hydrogen. The overall efficiency of producing hydrogen and then burning the hydrogen for steam power production is presently low, perhaps not more than 20 to 25 percent. Although the overall efficiency may be improved somewhat, it does not appear that hydrogen will be used extensively in this manner, at least not in the near future.

A considerable amount of effort is being expended on the use of hydrogen as a fuel for internal combustion engines, particularly in engines for cars, trucks and buses. The technical feasibility of this use of hydrogen has been demonstrated by several investigators. With the exception of nitrogen oxides, the exhaust from hydrogen burning engines is substantially pollution free. The amounts of nitrogen oxides produced in the hydrogen engine are much lower than for other fuels. It has been reported that comparable torques and power have been obtained in the hydrogen engine and, in some cases, higher efficiencies.

A major deterrent to the use of hydrogen in internal combustion engines is the problem of how to carry the hydrogen. Even when it is highly compressed its volume is still too bulky to permit a sufficient amount of it to be carried in automobiles. Furthermore, the tank must be very heavy to stand the high pressures. It is possible to liquefy hydrogen by cooling it to very low temperatures and subjecting it to high pressures. (Hydrogen must be cooled to below 60°R to liquefy it). To keep it in the liquid state, the hydrogen would have to be carried in a special thermos-like (cryogenic) tank, which is bulky and very heavy. Thus, it does not appear to be feasible to use liquid hydrogen for automotive service.

A promising alternative method of transporting hydrogen for automotive purposes is in the use of metal hydrides. Because of its high mobility and its small molecular size, gaseous hydrogen can penetrate the crystal structure of some metals and their alloys. For some metals, particularly titanium, the penetration is so great that the amount of hydrogen present in a given volume of the hydride exceeds that of liquid hydrogen. A tank containing these hydrides can be charged by introducing hydrogen under pressure. For some of the hydrides, hydrogen will be released if the pressure on them is reduced sufficiently. Other hydrides require heating (for which the engine exhaust may be utilized) to reject the hydrogen. Unfortunately, most of the hydrides, which are capable of absorbing large amounts of hydrogen, are at present, too expensive for this class of service. It is not possible at this time to predict whether or not newer hydrides will be developed at sufficiently low costs to permit this method of hydrogen storage for automotive service to be practical.

The situation is somewhat different when considering hydrogen for aircraft fuel. This use of hydrogen has been studied by NASA. It appears that the storage of liquid hydrogen in cryogenic tanks for aircraft engines may be feasible in spite of the large weight of the cryogenic tanks. The much larger heating value of hydrogen relative to that of other liquid fuels on the pound basis means that the combined weight of the hydrogen and its tank is materially less than that of other fuels and their tanks for a given energy content. However, because of its low density, the liquid hydrogen fuel tanks would be somewhat larger than conventional tanks and hence may offer added wind resistance. Particularly for supersonic planes, liquid hydrogen fueling has an additional advantage. The cold hydrogen on its way from the tanks to the engine can be used to provide some cooling of parts of the plane which tend to overheat. Information is lacking to predict the extent of the use of hydrogen as an aircraft fuel. Even with increasing costs of conventional fuels, it appears that there must be a large reduction in the cost of hydrogen before it will be used extensively for this purpose. Ultimately, however, with the growing shortages of petroleum products we have to consider the use of hydrogen for internal combustion power regardless of its cost, particularly in the case of an oil embargo.

21.6 SUMMARY ON THE HYDROGEN ECONOMY

Hydrogen is not a primary fuel. Energy from another source must be used to produce it, with the result that its energy is less than the energy required to produce it. Its value lies in its ability to store energy, to transport energy, and in its potential to supplant oil and natural gas for those types of services for which they are most suited.

Hydrogen can be produced by thermo-chemical processes in which it is released by chemical reactions, with the energy required to produce the hydrogen coming from heat which is supplied. This method has been demonstrated in the laboratory to be technically feasible. Large scale operations will be required to evaluate the economics of this process.

Hydrogen has been produced for many years by electrolysis of water. The efficiency of this process is relatively low but probably can be improved considerably. The initial cost of the equipment is high. A large increase in the amount of hydrogen produced by electrolysis can take place only if the efficiency can be improved, the cost of production lowered, and the cost of electrical power available for this purpose can be kept low.

In addition to its many present uses in industry, hydrogen has the potential for helping to alleviate our energy problems by providing energy storage capabilities, by offering a potentially good method for energy transportation, and by substituting for natural gas for heating and for gasoline for internal combustion uses. It also has the potential for the production of large amounts of electric power when used in the hydrogen-oxygen fuel cell, particularly when electrical loads are widely distributed.

In addition to being produced by electrolysis in conventional power plants at times of low power demands, it is proposed to generate it, also by electrolysis, at wind power plants, ocean thermal gradient power plants and, possibly, at solar power plants.

Hydrogen has much potential to help in using energy. In general, many problems, particularly costs, must be solved before hydrogen can reach its potential. Because information is lacking on large-scale use of hydrogen in energy-involved processes, it is very difficult to predict when and how widely hydrogen will be substituted for other methods of energy use. Certainly it does not appear that hydrogen will play an important role in the energy field in the next decade and perhaps in the next two decades.

SUPPLEMENT A

ILLUSTRATIVE EXAMPLES

Example 21-1. It is proposed to furnish power to a rural area by installing a 26,000 kW fuel cell plant, which is to operate on hydrogen. The plant capacity factor is 60 percent (i.e., the ratio of the average to the rated output). The hydrogen is to be produced by electrolysis in a fossil fuel power plant having an average efficiency of 38.5 percent. The average efficiency of the electrolysis plant is 62 percent and that of the fuel cells plant is 45 percent. Determine the tons of coal required per hour on the average, if the coal has a heating value of 13,100 Btu per pound.

Solution. The overall efficiency of the power plant, the electrolysis plant and the fuel cell plant combined equals

$$0.385 \times 0.62 \times 0.45 \times 100 = 10.74\%$$

The Btu delivered per pound of coal equals

$$13{,}100 \times 0.1074 = 1{,}407 \text{ Btu}$$

The average load equals

$$26{,}000 \times 0.60 = 15{,}600 \text{ kW}$$

The tons of coal required per hour equals

$$\frac{15{,}600 \times 3{,}413}{1{,}407 \times 2{,}000} = 18.92 \text{ tons} \qquad \text{answer}$$

Example 21-2. It is proposed to generate electric power by the use of off-shore wind-power plants and to use the electrical power thus produced to generate hydrogen by electrolysis of water. The hydrogen is to be piped to land and used for electric power generation. The average power output of the hydrogen fueled power plant is 200,000 kW and its efficiency is 36 percent. The efficiency of the electrolysis plant is 58 percent. Determine the average power output of the wind power plant.

Solution. The overall efficiency of the electrolysis plant and the hydrogen power plant equals

$$0.36 \times 0.58 \times 100 = 20.88\%$$

The average power output of the wind-power plant equals

$$\frac{200{,}000}{0.2088} = 957{,}900 \text{ kW} \qquad \text{answer}$$

Note: the actual output of the wind-power plant must be higher since the frictional losses in the hydrogen pipeline have been neglected.

Example 21-3. Determine the average number of cubic feet of hydrogen (standard conditions) which must be handled in Example 21-2. The heating value of the hydrogen at standard conditions is approximately 325 Btu per cubic foot.

Solution. The average energy input to the hydrogen power plant equals

$$\frac{200{,}000 \times 3{,}413}{0.36} = 1{,}896{,}000{,}000 \text{ Btu/hr}$$

$$= 31{,}600{,}000 \text{ Btu/min}$$

The amount of hydrogen required equals

$$\frac{31{,}600{,}000}{325} = 97{,}230 \text{ cu.ft./min.} \qquad \text{answer}$$

Example 21-4. At present, 1,000 gallons of oil are used per year for heating a house. Determine the number of cubic feet of hydrogen (standard conditions) if it is to replace the oil. Take the heating value of the oil to be 142,000 Btu per gallon and the heating value of the hydrogen as 325 Btu per standard cubic foot. Assume also that the efficiencies of burning the two fuels are equal.

Solution. The heating value of the oil used per year equals

$$1{,}000 \times 142{,}000 = 142{,}000{,}000 \text{ Btu}$$

The amount of hydrogen required per year equals

$$\frac{142{,}000{,}000}{325} = 436{,}900 \text{ cu. ft./yr.} \qquad \text{answer}$$

SUPPLEMENT B

PROBLEMS

21-1. A rural area has need of an average of 10,000 kW of electrical energy. It is proposed to produce hydrogen by electrolysis and pipe it to the point of use and then to produce power in a small steam power plant. The average efficiency of the small power plant is 30 percent. The central station plant producing the hydrogen has an average efficiency of 38 percent. The efficiency of the electrolysis plant is 60 percent. Determine the tons of coal required in the central station plant if the coal has a heating value of 12,800 Btu per pound.

21-2. Calculate the overall efficiency for Problem 21-1.

21-3. Discuss the desirability of the method given in Problem 21-1 for the production of the 10,000 kW.

21-4. Particularly if the cost of producing hydrogen can be reduced, would it be more desirable to use the hydrogen in the rural area of Problem 21-1 for residential heating rather than power production?

21-5. Assuming the fuel cells can be developed at a sufficiently low cost, discuss the desirability of using fuel cells in Problem 21-1 rather than a steam power plant.

21-6. At standard conditions of 14.7 psia, 77°F, hydrogen has a heating value of approximately 325 Btu per cubic foot. Determine the number of cubic feet of hydrogen which has the same heating value as a tank of 20 gallons of gasoline. Assume the gasoline has a heating value of 145,000 Btu per gallon.

21-7. Assuming that the heating value of hydrogen varies directly as its pressure, determine the pressure required to reduce the volume of the hydrogen in Problem 21-6 to that of a 20 gallon gasoline tank. (There are 231 cu. in. per gallon.)

21-8. Discuss the desirability of trying to develop satisfactory methods of transmitting electrical energy ashore from ocean thermal gradient power plants rather than to produce hydrogen, pipe it ashore and then use it to produce power in steam power plants.

21-9. The energy developed in an ocean thermal gradient power plant is to be used ashore for heating purposes rather than for electric generation as in Problem 21-8. Discuss the desirability of producing hydrogen for this purpose rather than transmitting the electrical energy ashore.

21-10. a. As an alternative to using fuel cells for power generation in Example 21-1, it is proposed to make the 26,000 kW power plant a coal burning plant. Determine the tons of coal required per hour if the plant efficiency is 28 percent.

b. Do you recommend the coal burning power plant?

22

New Conversion Methods

22.1 INTRODUCTION

As has been stated earlier, present indications are that electrical sources will furnish us with an ever-increasing portion of our energy. Neglecting the relatively small percentage of electrical energy which is generated from hydropower, fossil and nuclear fuels presently are the source of our electrical energy. Due to limitations in the transformation of heat into electrical energy, the maximum efficiency for such a conversion today is about 40 percent. Because of this low efficiency much effort is being expended to try to obtain higher efficiencies and thus to conserve our fuel resources.

As discussed previously also, it is possible to attain higher efficiencies in steam power plants by increasing steam temperatures by two hundred degrees or so from the present temperatures of around 1,000°F. We have the materials to stand this temperature but they are so expensive that their use in steam power plants would increase the cost of power, in spite of the higher efficiencies attainable with their use. Although government subsidies would permit their use and thus conserve fuel, such an action is not desirable, since the high temperature materials are in short supply and must, for the most part, be imported. As an alternative to using higher steam temperatures, use of so-called dual cycles is being considered.

At present, power is produced from fossil fuel by first producing mechanical power (e.g., rotation of a turbine wheel) and, from that, electrical power. Direct production of electrical power eliminates the need for developing mechanical power and hence has the potential for higher overall efficiencies. The four most common direct conversion methods now under consideration are

a. Thermoelectric conversion
b. Thermionic conversion
c. Magnetohydrodynamics (MHD)
d. Fuel cells.

Thermoelectric* and thermionic† conversions show little promise for large-scale power production and hence will not be discussed here. MHD and fuel cells show promise for the production of power on the large-scale basis. They will be discussed in following sections.

22.2 DUAL CYCLES

In a steam power plant, water normally enters the steam generator (boiler) at a temperature considerably below its boiling temperature. The water is heated to its boiling temperature, boiled, and then the steam produced is heated much higher than the boiling temperature. The mean temperature at which heat is added is much lower than the final steam temperature. Fifty years ago, pressures around 400 to 600 psia were common for steam power plants. Hence the boiling temperatures were relatively low. To improve the efficiency some mercury vapor plants were developed. Mercury boils at a relatively high temperature at moderate pressures, hence permitting heat to be added at high temperatures. But mercury, as it rejects heat, condenses at relatively high temperatures. To establish a large overall temperature range, the heat given up by the condensing mercury vapor was used to produce steam for use in a steam turbine. This arrangement is shown in Figure 22-1. Since the steam condenses at a low temperature, the overall temperature range and efficiency was high. Mercury, however, was expensive and limited in supply. Many problems were encountered in its use. As steam pressures and resulting

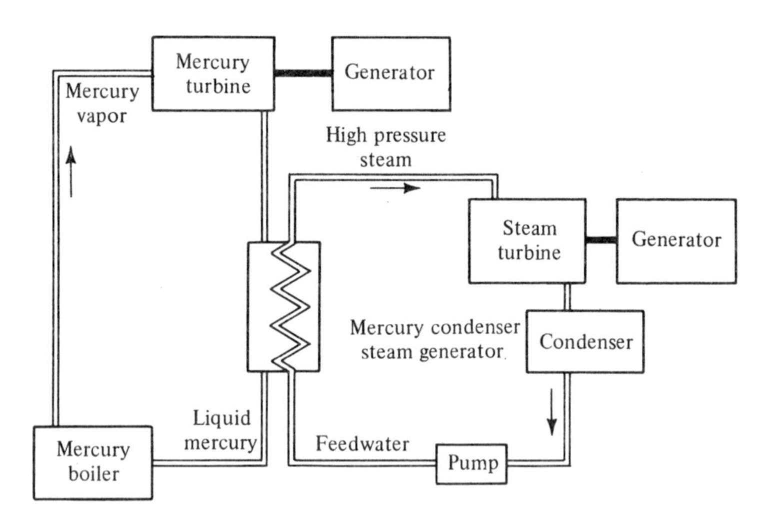

Figure 22-1. Binary vapor power plant

* When strips of two dissimilar conductors are joined at their ends and one junction heated, a voltage is created, thus offering the opportunity for power delivery.

† When a metal is heated to very high temperatures, electrons are given off from the free surface. The electrons can be collected and channeled as power to an external load.

boiling temperatures were increased, efficiencies of the conventional steam power plant approached that of the mercury vapor-steam power plant. Consequently, the mercury vapor plants went out of existence. Today, some consideration is being given to reviving the dual vapor cycles. Other liquid metals, such as rubidium, offer a greater potential since they boil at higher temperatures than does mercury. Since even pilot plants using these vapors have not been built, it is not possible to evaluate the feasibility of their use on a large-scale basis.

Much more effort is now being expended in the development of the gas turbine-steam turbine power plant. This combination has been used by several power companies in connection with older, low pressure plants. It was found that after many years of operation, the steam generator (boiler) had deteriorated so much that it was unusuable. The steam turbine and condenser were still serviceable. Rather than replace the steam generator, a power-producing gas turbine unit was installed, with the hot exhaust gases from the gas turbine generating steam for the old steam turbine. (See Figure 22-2.) Not only was the total power output increased, but the thermal efficiency of the two units was much higher than that of the old steam power plant. The overall efficiency, however, was not better than that of modern steam power plants.

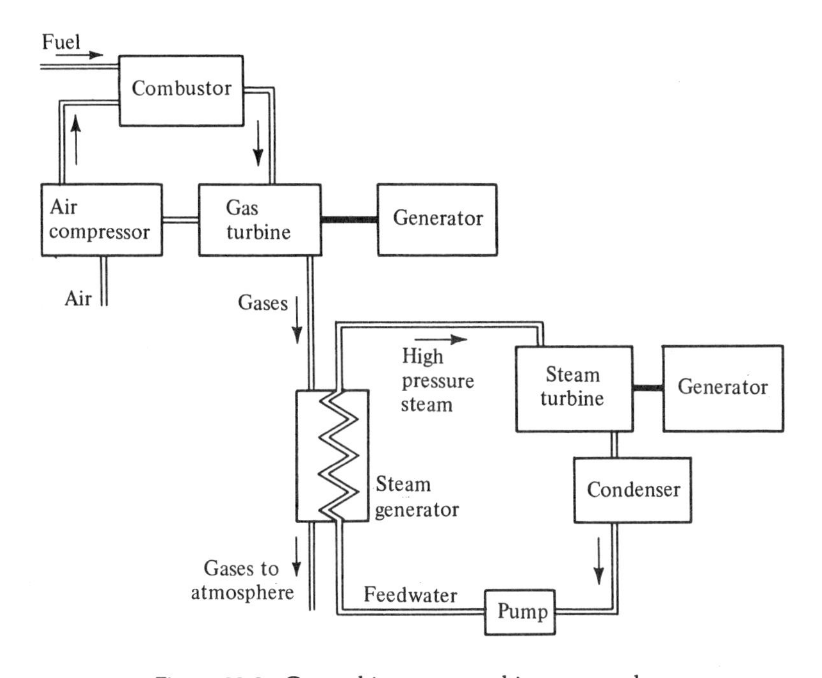

Figure 22-2. Gas turbine-steam turbine power plant

The overall efficiency of the gas turbine-steam turbine power plant is limited by the allowable operating temperatures for the gas turbine. Until rather recently, the allowable gas temperature for stationary units was around

1,600°F. This temperature is now being increased. Much investigational effort is being made to permit the use of gas temperatures of at least 2,200°F to 2,400°F. To operate at such high temperatures, development work is under way on entirely new blade materials, particularly the ceramics. As an alternative, if proper blade cooling can be achieved, then present blade materials will be satisfactory for the high temperatures.

It is being predicted by many working in the field that overall efficiencies of at least as high as 45 percent will be achieved for the combined steam-gas turbine units in the next decade. If this prediction is fulfilled, then at least 10 percent less fuel will be required for a given electrical output. One drawback to a rapid growth in the use of steam-gas turbine systems is the fact that the fuel required is either oil or natural gas. For at least thirty years, work has been carried out to develop coal-burning gas turbines. To date, these efforts have not been successful on full-size units and it does not appear that success will be achieved at least in the near future. Much thought now is being given to the production of low Btu gas from coal with the gas being used as a fuel for gas turbines. Since this idea has not been carried out in a full-size plant, it is difficult to predict its economic feasibility. With increasing cost of oil and natural gas and their growing shortages, it appears that the use of low Btu gas in gas turbines may become a reality for large units. As a result, a sizeable amount of our electric power may be produced by gas-steam turbine plants.

22.3 COGENERATION

Many industries need sizeable amounts of electric power and also need heat for process work and for building heating. In the past, except for the largest industries, power has been purchased and fossil fuels have been burned to supply the desired amount of heat. Because the temperature of steam remains constant as it condenses, provided its pressure is held constant, when heating is required at relatively low or moderate temperatures, it has been the practice to use a boiler to produce the steam. With cogeneration the steam is produced at a high pressure and passed through a steam turbine. The exhaust pressure is kept sufficiently high so that the turbine exhaust steam, in condensing, will supply heat at the desired temperature. For example, when the exhaust pressure is 5 psig, steam condenses at 228°F. For exhaust pressure of 100 psig, steam condenses 338°F. In the ideal case, except for losses in the boiler, almost all of the energy of the fuel is used in producing electric energy and in producing the desired heating. This is a contrast to the electric power companies which must reject perhaps sixty percent of the energy put in the steam to the condenser. Thus, thermodynamically speaking, cogeneration is a very desirable method of conserving energy.

However, many factors must be taken into consideration before deciding to install a cogeneration system:

1. There must be a balance between the demand for electricity and heating steam. Unless most of the exhaust steam can be used for heating purposes, cogeneration may not be a good idea.
2. Small turbo-generators are very expensive per kilowatt rating. The operating labor costs are almost as large for a small unit as for a large unit. Cogeneration is questionable unless turbo-generator units of at least one or two thousand kilowatts are involved.
3. The maximum demands for steam and electrical energy must be considered. Do they occur at the same time?

Today many industries are giving serious consideration to installing cogeneration in an effort not only to reduce costs of operation but also to conserve energy. As an alternative to the use of steam turbines, particularly small ones, for cogeneration, some thought is now being given to using either gas turbines or diesel engines together with electric generators to produce electrical energy. The hot exhaust gases from these units can be used to produce steam required for heating.

22.3 MAGNETOHYDRODYNAMICS (MHD)

As discussed earlier, an ionized gas is produced when some of the electrons are freed from their atoms, leaving the atom as a positively charged positive ion. When an ionized gas is passed through a strong magnetic field, the electrons have a tendency to be diverted to an electrode at the side of the flow passage. (See Figure 22-3.) If a circuit is provided, the electrons will flow to an external load, thus delivering electrical energy. The positive ions being very heavy, are not significantly diverted by the magnetic field.

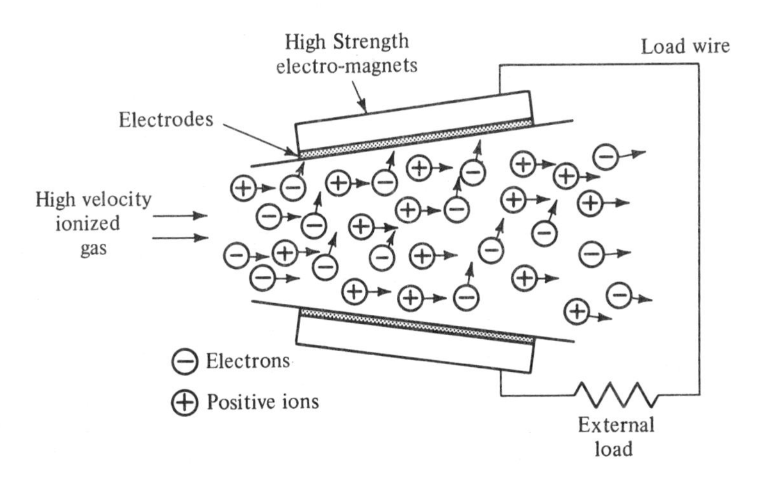

Figure 22-3. Elements of a MHD generator

Most gases, such as air, do not ionize appreciably even when heated to extremely high temperatures. Thus, to produce the desired ionization, a "seed" material, such as cesium, potassium or some of their compounds, is added. Even with the use of seeding material, the gas temperature entering the MHD generator is very high, perhaps 4,300°F to 4,500°F when cesium is the seed material and about 300°F for potassium. To achieve the desired temperatures, highly preheated air, say at least 3,000°F, is supplied to the combustor. An alternative is to use oxygen in place of air. When oxygen is used to burn fuels, much higher temperatures are produced since there is no nitrogen present to absorb much of the heat. The use of oxygen, however, will greatly increase the cost of power production. The cost of oxygen can be reduced if some of the electrical energy of the plant is used to produce hydrogen by electrolysis of water. In such a case, oxygen would be a by-product and, as such, should be relatively cheap.

In order to produce large amounts of power in a relatively small unit, gas velocities in a MHD generator must be high, perhaps in the order of at least a mile per second. To achieve these velocities, the gases are expanded in a nozzle before entering the MHD generator. This requires operating the combustor at a pressure of several atmospheres. Since the temperature of the gases leaving the MHD generator is high, perhaps 3,500° to 4,000°F, these gases can be used to preheat the air to the required temperature of at least 3,000°F before its entrance into the combustor. (See Figure 22-4.)

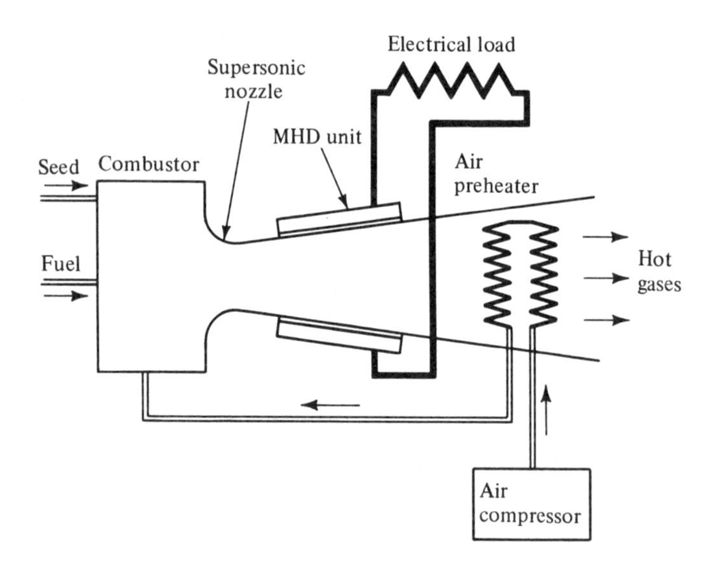

Figure 22-4. Elements of a MHD plant

Even after preheating the combustion air, the temperature of the exhaust gases are high, perhaps between 2,000° and 2,500°F. To conserve the energy of these gases, they are used to produce steam at temperatures and pressures comparable to those of modern conventional steam power plants. A simplified line diagram is shown in Figure 22-5 for the combined MHD-steam power plant. Thermal efficiencies ranging from 50 to 55 percent have been predicted for the combined MHD-steam power plants. If these efficiencies can be attained, then at least 25 percent less fuel will be required for a given electrical output. Thus the MHD generator has the potential of making a major contribution toward conservation of energy.

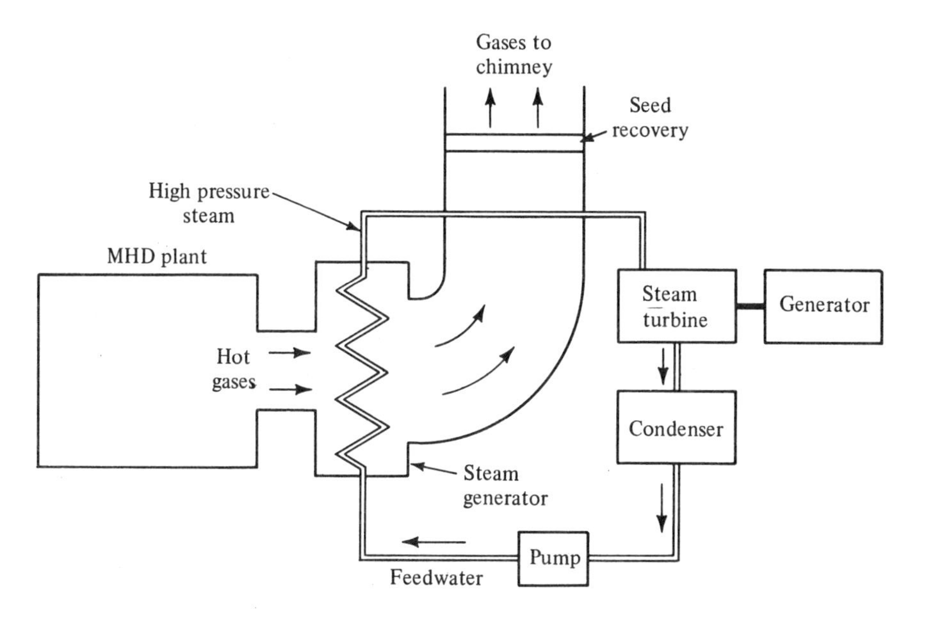

Figure 22-5. Elements of a steam power plant-MHD plant

Unfortunately, there are many major problems which must be solved before the MHD generator can become a commercial reality, such as

1. Preheating the air
2. Producing electrodes to withstand the operating conditions
3. Minimizing the power for the electromagnets
4. Recovery of seed material
5. Minimizing the nitrogen oxides normally formed

Preheating the air is accomplished in a heat exchanger, using the hot gases leaving the MHD generator. Because of the temperatures involved, no material presently is completely satisfactory for this purpose. Steels will soften and then melt at temperatures below that required for preheating the air. Although some ceramic materials will stand the temperatures, normally these materials do not possess the high thermal conductivity required for good heat exchanger performance.

Particularly when coal is used as the fuel, the ash-laden, very high velocity hot gases tend to erode the electrodes unless they are cooled. Cooling the electrodes may cause the molten ash to solidify on them, destroying their electrical conductivity.

Under normal conditions, much power is required to operate the electromagnets. This power can be greatly reduced by using super conductors. However, this requires operating the magnets at a temperature only slightly above absolute zero. This problem is greatly complicated by the fact that the electromagnets are close to that part of the generator where the gas temperature is over 4,000°F.

The seed material is very expensive. To hold the cost of the seed material down to 5 percent of that of the coal used, approximately 99 percent of the seed material must be recovered when potassium is used and a higher percentage when cesium is the seed material.

Because of the very high temperatures involved, perhaps 10 times as much nitrogen oxides are formed in a MHD system as in the conventional fossil fuel power plant. This amount of nitrogen oxides is far in excess of allowable standards.

The problems listed above are very difficult ones. It is very hard to predict when they can be solved. Certainly, their solutions will add to the cost of power production in MHD generators. However, the potential for large savings in fuel justifies the extensive effort being expended to make MHD power generation a commercial reality.

22.5 FUEL CELLS

In certain respects, fuel cells resemble electric storage batteries. In batteries, however, there is depletion of some of the materials. After a period of time, it becomes necessary to either replace these materials or restore them to their original condition by charging the battery. On the other hand, in fuel cells, the reactants are supplied continuously and the products of the reactants are removed as they are formed. In both devices there is a direct conversion of chemical energy into electrical energy which is delivered by electrons to an external load.

The concept of the fuel cell is an old one. In 1839, Sir William Grove built a hydrogen-oxygen fuel cell. Since that time development work on fuel cells moved ahead rather slowly until the late 1950's when fuel cell development expanded, with many hundreds of millions of dollars being spent. One result of this development is the hydrogen-oxygen fuel cell which has been used extensively to provide auxiliary power for spacecraft uses.

There are several types of fuel cells, but the most common one is the hydrogen-oxygen cell. A line diagram of this cell is shown in Figure 22-6. Fundamentally, the cell consists of two electrodes separated by an electrolyte. Hydrogen is supplied to the fuel electrode. By catalytic action of the electrode,

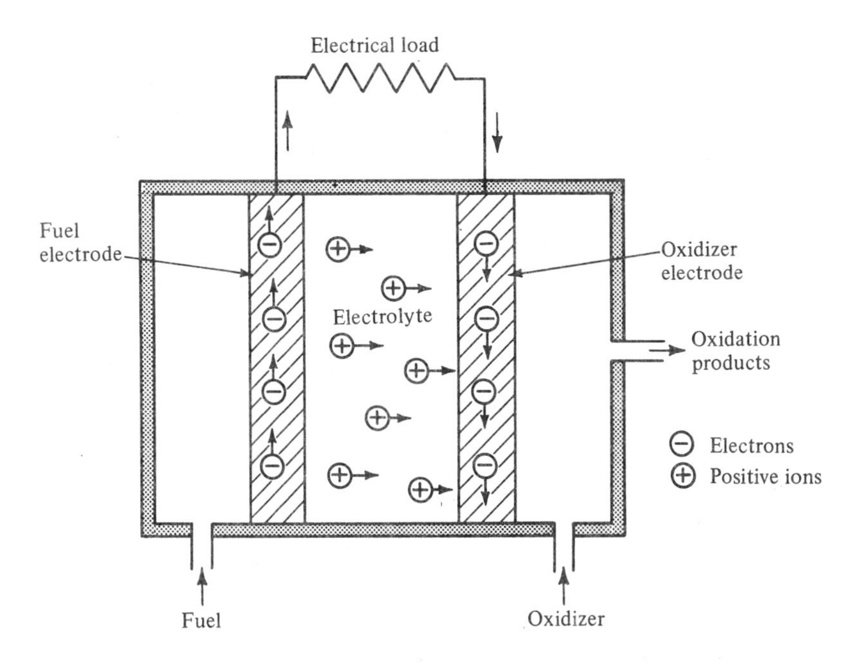

Figure 22-6. Elements of a fuel cell

two molecules of hydrogen are ionized into four positive ions and four electrons

$$2H_2 \rightarrow 4H^+ + 4e^- \quad (22\text{-}1)$$

where H^+ designates positive hydrogen ions and e^- the negative electrons.

The electrons pass through the external circuit, thus delivering electrical energy. The positive hydrogen ions pass through the electrolyte to the oxygen electrode. Here they combine

$$4e^- + O_2 + 4H^+ \rightarrow 2H_2O \quad (22\text{-}2)$$

The overall net result of the combined action at the two electrodes can be obtained by adding the two equations.

$$2H_2 + O_2 \rightarrow 2H_2O \quad (22\text{-}3)$$

It should be noted that although the overall result is the same as that for burning hydrogen, the hydrogen and oxygen molecules actually do not come in direct contact with each other in the fuel cell. Using other electrodes and electrolytes, different actions can occur. At the fuel electrode,

$$2H_2 + 4OH^- \rightarrow 4H_2O + 4e^- \quad (22\text{-}4)$$

and at the oxygen electrode,

$$2H_2O + 4e^- + O_2 \rightarrow 4OH^- \quad (22\text{-}5)$$

where OH^- designates the negative hydroxyl ions. The OH^- ions pass through the electrolyte over to the fuel electrode to combine with the hydrogen.

The actual fuel cell is not as simple as it may seem. The fuel electrode must provide the proper catalytic action. It must be porous enough so that ample surface is provided for the hydrogen to be ionized. The oxygen electrode must also be porous. Both electrodes must permit ready flow of electrons. The electrolyte, on the other hand, must resist the passage of electrons but must readily conduct the positive hydrogen ions. To ensure good performance, the fuel cell may have to be operated at elevated temperatures and pressures. Both the electrodes and electrolyte must be stable and non-deteriorating when functioning under normal operating conditions.

The conventional power producing devices and also the direct energy converters, such as the thermionic, the thermoelectric and the MHD generators are heat engines. As such their ability to transform heat into work is limited by the temperature range under which they operate. The fuel cell is not a heat engine. It produces electrical energy directly from chemical energy. Hence, there is no temperature restriction on its energy conversion efficiency. Although the present efficiencies of fuel cells are comparable with those of fossil fuel plants, it seems reasonable to expect the efficiencies to reach at least as high as 50 percent. It is difficult to predict how much more the efficiencies of fuel cells can be increased in the future. It can be readily shown that the theoretical efficiencies of certain fuel cells can exceed 80 percent.

Because of their potential, some of the larger utility manufacturing companies such as General Electric and Westinghouse have devoted a substantial amount of their resources to the development of fuel cells for commercial purposes. Very extensive work is underway at Power Systems Division of United Technologies Corporation (formerly the Pratt and Whitney Division of the United Aircraft Corporation). This work has grown out of the production of fuel cells for auxiliary power for spacecraft. With support from many gas utility companies, United Technologies is developing fuel cells in units ranging from 10 to 250 kW. These units, using natural gas, are to be located in residences, commercial buildings and small industries, to supply all of the power and, in some cases, much of the heat. A few 12 kW fuel cells have been placed in residences.

With the support of nine power companies, United Technologies is also developing 26,000 kW fuel cell power modules. The overall development of fuel cells is now being supported by the major electric utilities, the National Rural Electric Cooperative Association, the American Public Power Association, the American Gas Association, the Edison Electric Institute, the Electric Power Research Institute, and the federal government. With such strong support, it should be expected that many of the major problems relative to fuel cells will be solved and that extensive use of fuel cells will take place in the next decade or two.

In addition to the technological problems of the fuel cell itself, one of the major problems involved with fuel cells is that of the fuel. It has been demonstrated that hydrogen is a very satisfactory fuel for fuel cells. However, as pointed out in Chapter 21, hydrogen is not a primary fuel but is obtained by the expenditure of some form of prime energy. It does not seem logical to generate electrical energy in a conventional manner to produce hydrogen from water and then use hydrogen to produce electrical energy. However, it may become desirable to do this in special cases if the efficiency of both the production of hydrogen and its use in the production of electrical energy can be increased significantly. If power is to be produced at some distance offshore by either windmill plants or by OTEC plants it may be desirable to produce hydrogen and pipe it ashore. It then could be used either for heating or for power production. If fuel cells can be developed which are reasonable in cost and have efficiencies much higher than at present, this may be very desirable.

By reforming it (i.e., changing its nature), natural gas can be used in fuel cells. However, natural gas supplies are decreasing, hence natural gas does not offer much potential for large-scale future use in fuel cells. Pipeline gas, produced from coal, may be substituted for natural gas provided its cost can be reduced sufficiently to justify its use in fuel cells. Some attempt has been made, particularly by Westinghouse Corporation, to generate industrial gas and to use it in fuel cells (i.e., the hydrogen and carbon monoxide in the gas). It has not been demonstrated as yet that this method is commercially feasible. Some work is now underway to reform oil, such as Number 2 oil, to provide a fuel for fuel cells. It is true that oil is domestically in short supply but much of our electric power is now produced by the use of oil, and it is projected that this will continue for many years. If fuel cells can be developed to use oil efficiently, there will be significant reduction in the amount of oil required for power production.

A second major problem relative to fuel cells is that of costs. Presently fuel cell costs are materially higher than those of conventional energy conversion devices. It is felt, however, that costs can be reduced sufficiently to make it possible to produce power from fuels at comparable or even lower costs than by other methods. This may become true when fuel cells are produced on a large-scale basis. The potential for reducing the costs of fuel cells may be evidenced by the various large groups now supporting its development.

In addition to the potential for producing power at higher efficiencies and thus conserving fuel, and also of producing power at lower costs, the fuel cell has other advantages. If hydrogen is the fuel, the fuel cell is nonpolluting, adding only heat and water vapor to the atmosphere. If the fuel cell obtains its fuel by reforming other fuels, pollution occurs only in the reforming process. Normally, this can be controlled satisfactorily.

As stated earlier, it is desirable for conventional power generation to use very large, efficient units, with at least two or three units grouped together. This means that electrical energy must be transmitted over considerable distances, particularly when the load is not predominantly an urban one. The costs of the long transmission lines together with high line losses can be avoided by the use of fuel cells. Since the fuel cell modules are composed of many small individual cells, the cost and the efficiency of the fuel cell modules are almost independent of size. This permits them to be located at stations close to the point of demand. The fuel cells, being almost completely self-contained, can be operated remotely.

22.6 SUMMARY ON NEW CONVERSION METHODS

Various methods have been proposed for obtaining efficiencies materially higher than the efficiencies presently obtainable in conventional power generating plants. Three of these proposed methods show much promise. These methods are the combined gas turbine-steam turbine plant, the magnetohydrodynamics-steam turbine plant, and the fuel cell plant. All three methods offer the potential of producing a given amount of power for perhaps 20 to 30 percent less fuel than is now possible in conventional power plants. If this becomes a reality it will affect our future energy requirements significantly since it has been predicted that we will use approximately 40 percent of our total energy supply for electric power generation before the end of the century.

There are many problems to be solved before these three methods of power production can be used extensively. One problem is that these three methods, especially the fuel cell, require special fuels. Many technical problems, particularly for the magnetohydrodynamics generator, are presently major obstacles to the uses of these methods. Costs of these three methods when developed to meet their potentials must be reduced from present indicated values. The potential for these three methods of power production is sufficiently large to justify the resources presently being devoted to their development. Most of those working in the field feel that the potential of each of these three methods will be realized in the next decade or two.

SUPPLEMENT A

ILLUSTRATIVE EXAMPLES

Example 22-1. Calculate the tons of circulating water required per second for the condensers of a conventional power plant having an average output of 1,800,000 kW and an average efficiency of 38 percent. The temperature rise of the water is 16°F. Assume that 10 percent of the heat supplied is lost as heat from the system and the remaining 90 percent is either delivered as work or is removed in the condenser.

Solution. Per kW output, the heat supplied equals

$$\frac{3{,}413}{0.38} = 8{,}980 \text{ Btu per hour}$$

The heat rejected to the condenser per kW output equals

$$8{,}980 \times 0.9 - 3{,}413 = 4{,}670 \text{ Btu per hour or } 1.297 \text{ Btu per second.}$$

The total heat rejected to the condenser equals

$$1.297 \times 1{,}800{,}000 = 2{,}335{,}000 \text{ Btu per second}$$

The circulating water required equals

$$\frac{2{,}335{,}000}{16 \times 2{,}000} = 72.98 \text{ tons per second} \qquad \text{answer}$$

Example 22-2. Same as Example 22-1 except a MHD-steam plant having an efficiency of 50 percent is used.

Solution. The heat supplied per kW output equals

$$\frac{3{,}413}{0.5} = 6{,}826 \text{ Btu per hour.}$$

The heat rejected to the condensers per kW output equals

$$6{,}826 \times 0.9 - 3{,}413 = 2{,}730 \text{ Btu per hour or } 0.7854 \text{ Btu per second.}$$

The total heat rejected to the condensers equals

$$0.7854 \times 1{,}800{,}000 = 1{,}414{,}000 \text{ Btu per second}$$

The circulating water required equals

$$\frac{1{,}414{,}000}{16 \times 2{,}000} = 44.18 \text{ tons per second} \qquad \text{answer}$$

Note: A comparison of the answers of Example 22-1 and 22-2 shows a savings of 28.8 tons of circulating water per second when using the MHD plant.

Example 22-3. A MHD power plant is to burn 450 tons of coal per hour. There are 12.5 pounds of air per pound of coal. The air for burning the coal is to be heated from 600°F to 3,000°F before entering the MHD combustion chamber. Determine the amount of heat to be added to the air per hour. Take the value of c_p of the air as 0.28 Btu/lb.-°F.

Solution. The pounds of air to be heated per hour equals

$$450 \times 2{,}000 \times 12.5 = 11{,}250{,}000 \text{ pounds}$$

The heat required per pound of air equals

$$1 \times 0.28(3{,}000 - 600) = 672 \text{ Btu per pound}$$

The total heat required equals

$$11{,}250{,}000 \times 672 = 7{,}560{,}000{,}000 \text{ Btu/hour} \qquad \text{answer}$$

Note: This tremendous amount of heat must be transferred in the heat exchanger (air preheater).

Example 22-4. The power plants in Example 22-1 and 22-2 burn coal having a heating value of 13,100 Btu per pound and costing 40 dollars per ton. Calculate

a. The tons of coal required per hour for each plant.
b. The reduction in the cost of coal per year when the MHD plant is used.

Solution. For the conventional power plant, from Example 22-1, the heat supplied per kWh is 8,980 Btu. The tons of coal required per hour equals

$$\frac{1{,}800{,}000 \times 8{,}980}{13{,}100 \times 2{,}000} = 616.9 \text{ tons}$$

For the MHD-steam plant, from Example 22-2, the heat supplied per kWh is 6,862 Btu.

The tons of coal required per hour equals

$$\frac{1{,}800{,}000 \times 6{,}862}{13{,}100 \times 2{,}000} = 471.4 \text{ tons}$$

$$\text{The reduction} = 616.9 - 471.4 = 145.5 \text{ tons per hour.}$$

The reduction in cost of coal per year equals

$$145.5 \times 40 \times 24 \times 365 = \$50{,}980{,}000 \text{ per year} \qquad \text{answer}$$

SUPPLEMENT B

PROBLEMS

22-1. A 3,000,000 kW steam power plant now burns oil having a heating value of 5,800,000 Btu per barrel. The average load is 2,150,000 kW and the average efficiency is 38 percent. It is anticipated that a MHD-steam power plant will have an average efficiency of 50 percent. Determine the number of barrels of oil per day which can be saved if a MHD-steam power plant can be used.

22-2. Same as Problem 22-1 except a gas turbine-steam turbine plant having an average efficiency of 45 percent is to be used in place of the conventional steam power plant.

22-3. A fuel cell power plant, using hydrogen, is to be substituted for the conventional plant of Problem 22-1. If the hydrogen has a heating value of 325 Btu per cubic foot under standard conditions, determine the cubic feet of hydrogen required per day, assuming that the fuel cell efficiency is 46 percent.

22-4. Will it be necessary to reduce the initial cost per kilowatt rating of the three new conversion devices discussed here to approximately equal that of conventional power plants before they can be considered for use for commercial power generation? Why?

22-5. Discuss the desirability of using a large portion of our total resources to develop all of the three energy conversion methods discussed here, on one or two of them or on none of them.

22-6. A small industry has a somewhat constant electrical load of about 400 kW throughout the day. If this electrical energy is produced in a steam power plant, approximately 12,000 pounds per hour of turbine exhaust steam can be produced for heating purposes. The industry has need of approximately 4,000 pounds of steam per hour for three hours a day. Should cogeneration be used?

22-7. Discuss the possibility of using the three new conversion methods for automotive service.

PART V

CONSERVATION AND THE ENVIRONMENT

23

Energy Conservation

23.1 INTRODUCTION

Gifford Pinchot, a leading figure in the establishment of the conservation movement in this country, stated "Conservation means the wise use of the earth and its resources." This statement is sufficiently broad to include all efforts made for conserving energy. The difficulty arises in understanding the full implication of the word "wise." Certainly it is unwise to leave lights turned on in rooms that are not being used. It is unwise to leave air conditioners running in homes while the families are on vacation. But how can we decide if it is wise to use energy intensive machines to supplant hand labor, particularly if the hand labor requirements to perform the given tasks are not very exhausting? In such cases is it wise to substitute machines for hand labor, thus causing additional unemployment?

There are two extremes in the use of energy. On the one hand, there are many tasks which now are performed by energy demanding machines which are very difficult or almost impossible to perform with hand labor. As examples of such tasks, consider stamping out of parts for cars and machinery, raising an elevator, or moving a truck. Here the use of energy is almost mandatory. On the other hand, there are tasks for which we may, out of habit or custom, use machines, where these tasks may be performed quite readily by hand. In between these extremes are the vast number of energy uses which should be judged individually. In all cases, it should be asked whether or not it is desirable to use an energy consuming device. If the answer is positive, then the question should be asked whether or not the device is using the minimum amount of energy necessary to fulfill the stated purpose of the device. A seconddary question to be asked is whether or not the device is using energy which is in short supply or whether energy in a more readily available form can be used satisfactorily.

These are specific questions to be asked for specific cases and call for specific answers. But a broader question must be answered first. How much are we willing to pay for energy? As has been shown throughout this book, energy will become available to meet all anticipated demands if the public is willing to pay a sufficiently high enough price and if adverse environmental effects of procuring this energy are disregarded. The adverse environmental effects can be minimized but only at a price. For example, coal can be obtained by strip-mining and the land restored to satisfactory conditions if the public is willing to pay for the added costs.

To meet the ever-increasing demands, energy can be obtained from such sources as solar energy, wind energy, and ocean thermal gradient energy. But the public must be prepared to pay very high prices for this energy. Foreign oil may be purchased in increasing amounts, thus increasing our unfavorable trade balances. In short, if this country demands unrestricted supplies of energy, it is easily conceivable that, sooner or later, economic chaos will result.

We as a nation can prevent the very large future increases in the cost of energy only if our energy demands are restricted. It may be possible to hold energy prices close to the present level if energy demands were reduced to those of a decade or two ago. To accomplish such a reduction, it will be necessary to avoid using energy in many situations by shifting work loads from machines to the backs of people in factories, on farms, in transportation, and in homes. Furthermore, when energy is to be used, it will have to be used wisely. Thus the future price as well as the availability of energy will be governed to a large extent by our conservation efforts.

Tables 2-1, 2-2, 2-3 and 2-4 show the amount of energy now being used, the sources of this energy, and the purposes for which it is being used. The three major sectors which, directly or indirectly, consume most of our energy are industrial, transportation, and residential-commercial. A large portion of the total energy supplied is used for electric generation, but this energy, reduced by inherent losses in generation and transmission, is delivered to the three major energy using sectors. The methods of use of energy in these sectors differ in many respects and hence each sector will be discussed individually.

23.2 CONSERVATION IN ELECTRIC POWER GENERATION

Over the years, the electric power utilities, aided by the electric power equipment manufacturers, have made tremendous increases in the efficiency of power generation. Today, approximately three times the amount of electrical energy is produced from a given amount of fuel as was produced 75 years ago. Large gains in efficiency, however, should not be expected in the future, using present types of plants. As stated before, efficiencies can be improved by using higher steam temperatures. Although materials are available to permit operating at higher temperatures, they are so expensive that it will be more costly to produce power with their use, not only now but for many years.

A very small gain in efficiency can be expected by careful control of operating conditions. One reason for a relatively low overall efficiency of power plants is peak load demands. If highly efficient but expensive units were to be used for peak load conditions, fixed charges would be excessive. Hence, peak loads normally are carried by rather inefficient gas turbines or by low-efficiency semi-obsolete steam units. A partial solution to this problem is to use pumped storage, where feasible, to store energy to take care of peak load conditions (see Section 10.4). A better solution is to minimize peak loads. This can be done only by the customers. The customers can be encouraged, by a power rate structure, to avoid periods of high demands. This is relatively easy to do for industrial customers and is being done to some extent. It is more difficult to work with residential customers to avoid peak demands, particularly since their individual demands are relatively low and occur at various times throughout the day. However, a considerable amount of attention is now being devoted to peak pricing. It should be noted that the elimination of peak power demands will not only improve the generating efficiency but it will also decrease the need for additional power generating equipment, thus holding down fixed charges.

There is a good possibility that the three newer methods of power generation discussed in Chapter 22 (dual gas turbine-steam turbine cycles, magnetohydrodynamics-steam cycles, and fuel cells) will become sufficiently developed in the next few decades to economicaly use one or more of them for power production. Each of these three methods could result in a very significant saving of fuel. It is imperative that power company officials be fully cognizant of these developments and their potentials.

When electric power is to be transmitted long distances, line losses are significant. These losses are being reduced by designing the transmission lines to operate at increasingly high voltages. A promising method of reducing transmission line losses is to use superconducting materials. When their temperatures are reduced close to absolute zero, some conductors have such a low electrical resistance that line losses become very small. However, much energy is required to maintain the temperatures of the lines at the very low values. Because of the potential of this method of minimizing line losses, power company officials should also follow this development quite carefully.

23.3 CONSERVATION IN INDUSTRY

As may be seen from Table 2-2, approximately one-third of all the energy used in this country is by industry. There has been a rapid growth in the demand for energy by industry and this increase in demand is expected to continue. Much of this increase has been brought about by the shift from hand labor to machines. This shift has resulted not only in a large increase in the amount of energy required but also a decrease in the labor requirements for a given output of goods. This shift in work loads is a matter of national concern. In the future as this nation struggles to obtain the energy which its citizens demand, is it wise to increase the use of energy-intensive machines, both in the factories and outside them, which will displace workers and thus contribute to the unemployment problem? At present, the decision whether to continue the shift to larger and ever larger machines is made by factory managers almost solely on the basis of the economics of production.

In addition to the energy required for machine operation, industry uses much energy for heating. In all places in industry where energy is used, decisions must be made as to which form of energy is best suited for the task. For example, if the process calls for heating, is it desirable to use electricity for heating, to generate steam for the heating,* to burn natural gas, or to burn oil? In making the decision consideration must be given to the future availability of the energy which may be chosen. For example, natural gas may presently be the very best energy source for a given operation and also the least expensive. However, it must be recognized that in the not too distant future it will become very expensive and in short supply.

In addition to using machine processes only when really necessary, industry can save much energy by carefully examining two factors:

1. Is the energy used efficiently and effectively?
2. Can energy discharged in one operation be effectively used in a second operation?

The operations in industry are so varied that it is not possible to discuss these two factors in detail.

All too often in the past, the generation of steam for various uses in a plant was considered to be satisfactory as long as the steam was supplied when and where needed. In many cases, for example, when oil was burned to produce steam, insufficient attention was paid to burning the oil in an efficient manner. It has been estimated that in many installations, 10 to 15 percent of the fuel has been wasted because of improper operations of the furnaces. More energy has been lost in the steamlines by poor insulation and by leaks. Finally, as it was being used, in many cases steam has not been used wisely. As an example, the heat transfer equipment may not have been kept in the best operating condition.

In many industrial processes, waste heat must be removed. It is relatively easy to use water to remove the heat and then either to discard the water together with its energy or to use a cooling tower to cool the water for reuse. There may be a second process adjacent to the first one which requires heat to

* When very large quantities of heat are required, it may be desirable to produce steam at high pressures and pass it through a turbo-generator unit to generate electrical energy. The exhaust steam is excellent for heating, since its temperature remains constant as it condenses, providing that its pressure is held constant. (This is cogeneration (see Section 22-3).

be supplied to it at a relatively low temperature. Rather than to bring in energy from another source, it may be feasible to use the waste heat from the first process to supply the heat required in the second. Additional equipment may be required to use the waste energy but not only will energy be saved, but also the savings in many installations will pay for the additional equipment in a short period of time.

In practice, a close and continuing examination should be made of the planned installation of all energy consuming processes, the energy relationship between the various processes and also of their operation. Many industries, particularly larger ones, have made and are continuing to make such examinations. Energy savings up to 15 to 20 percent in one year have been reported and smaller but significant savings are being made in subsequent years. Not only have energy requirements been reduced but, because of the fuel savings, there has been a significant reduction in the cost of production. Although the larger industries generally have the engineering manpower available to make the desired savings in energy, many of the smaller industries do not have the engineering staff to handle this matter. Generally, it will pay these industries to bring in energy engineers to make detailed examinations of their energy uses. The potential for energy savings is sufficient to have a significant effect on our total energy demands. What is first required is for management of industries to realize that there is an opportunity not only to conserve the energy supply but also to reduce costs of operation.

There has been a radical change in the methods of farm operation in the last few decades. Many of the smaller farms have been absorbed by larger ones and the larger ones have grown. Except for those smaller farms still remaining, farm operation today is largely mechanized. In reality these larger farms are outdoor factories. As such, much of the discussion relative to energy conservation in industry applies equally well to farm operation in all its phases. The same type of decisions should be made, such as

1. Should machine labor supplant hand labor?
2. What type of machines should be used and what form of energy will be most satisfactory?
3. How can plant operations and equipment be designed for minimum energy use?
4. How can each piece of equipment be operated to use the least amount of energy?

23.4 CONSERVATION IN TRANSPORTATION

Approximately one-quarter of total energy supply is used for transportation, including the private cars, trucks, buses, trains, pipelines, boats and planes. The energy used for transportation is divided up approximately as follows:

	Percent of total transportation energy
Automobiles	52
Trucks	22
Aircraft	10
Military operations	5
Railroads	4
Ships and barges	4
Others	2
Buses	1

Almost two-thirds of the transportation energy is used for passenger transportation, with most of the remainder being used for freight. Thus, the greatest opportunity for energy conservation is passenger transportation. Since freight transportation consumes approximately eight percent of the total national energy supply, conservation in this area could also make a significant effect on our energy needs.

The least energy-demanding form of freight transportation is the pipeline. The pipeline is excellent for bulk transportation of liquids and gases. By mixing crushed coal and water a slurry can be formed which makes coal transportation by pipeline feasible. This has been done to a limited extent, particularly to transport coal to a large power plant. Some difficulties have been encountered securing the right-of-way for these pipelines. The railroads are said to object to coal pipelines using the railroad right-of-ways since the railroads depend on coal hauling for a sizeable portion of their operations.

Although waterway transportation of certain classes of freight requires less energy per ton mile than does rail transportation, in general these two modes require approximately the same amount of energy per ton mile of freight. Both of these modes require approximately 50 percent more energy per ton mile than does the pipeline. The energy required per ton mile to handle freight by trucks is approximately four times that for rail transportation. Truck transportation is convenient, flexible and more readily adaptable to handle relatively small quantities for short distances. Because of its high energy requirements, the use of truck transportation for hauling freight over long distances should be examined carefully.

By far the most energy demanding means of freight transportation is the airplane. It requires 50 to 60 times as much energy per ton mile as is used in rail transportation. Its use can be justified only in terms of convenience and speed of service, particularly for small packages.

Two factors should be considered in any attempt to conserve freight transportation energy. The owner of the goods to be transported should consider very carefully the various modes available to him. Energy requirements should be weighed very carefully against convenience. It should be recognized that, in general, the mode requiring the least amount of energy is also the cheapest. This will become increasingly true as energy costs escalate. Thus, it becomes a sound business practice to choose a low energy demanding mode for freight transportation.

The owner of transportation equipment has the opportunity to conserve energy. This equipment should be designed for low energy consumption, should be kept in good operating condition, and should be so operated that energy consumption is held to a minimum. Many companies, particularly the larger ones, carefully control these three factors and do operate with a low energy consumption. All too often, however, too many owners of transportation equipment are satisfied with their equipment if it runs satisfactorily and continues to do so with low maintenance cost, and fail to consider that there is an opportunity to conserve both energy and money.

There is a wide range of energy consumption per passenger mile for any given mode of transportation. This is particularly true for automobiles. Energy requirements vary with many factors, particularly the weight of the car and the type of service (urban or intercity). Also entering into the energy required per passenger mile is the number of passengers conveyed by a given vehicle. In general, however, for intercity service, rail transport requires approximately 75 percent more energy than buses, private automobiles over twice as much energy, and aircraft over five times as much energy per passenger mile as is required by buses. For urban transportation, the automobile uses on the average over twice as much energy per passenger mile as is used by the average of the various mass transit systems. Thus it is evident that citizens, by their choice of the mode for their transportation, can have a significant influence on the amount of energy to be used.

The average citizen has come to feel that it is his birthright to own a car and to operate that car whenever and wherever he or she pleases. There seems to be a feeling that the growing shortage of energy is a contrived one and price increases are totally unnecessary. Such attitudes must be changed if we are to avoid serious problems in the next decade or two. If a concerted effort is made by all citizens, it should be possible in passenger transportation to save three to four percent of the total energy we are now using for all purposes. Such a saving is possible without causing serious inconveniences.

Several factors are involved in the attempt to reduce the energy required for passenger transportation, assuming that cars are owned by most of the adult population:

1. Is the car selected with energy conservation in mind?
2. When should or should not the car be used?
3. If a car is to be used, how can its use be minimized?
4. Is the car in proper operating condition?
5. Is the car operated in a highly efficient manner?

It is well known that large, heavy cars having many cylinders are "gas guzzlers." For some years, consideration has been given as to how to drastically reduce the number of these large cars. When the Arab oil embargo was imposed and serious gasoline shortages occurred, there was a much larger demand for smaller cars, cars which used much less gasoline. However, as gasoline became plentiful again and seemed to be in ample supply, there was a decided shift in demand back to larger cars. Today, however, because of high gasoline costs and because federal requirements for increased miles per gallon, there has been a shift in demand back to smaller, high mileage cars. In general American manufacturers have not been fully prepared to produce well-built, high mileage small cars at a reasonable price. As a result, there has been a large increase in the sale of imported cars, particularly those from Japan. Because of the growing number of small cars and the high price of gasoline, there was a reduction in the total amount of gasoline used in this country in 1980 compared with 1979.

Much fuel can be saved by minimizing the use of cars. Some of the ways for a citizen to decrease the use of his or her car are

1. Carpooling.
2. Using public transportation.
3. Walking, where distances are short.
4. Using bicycles or motorcycles.
5. Avoiding "just riding around" by using the car only for necessary trips.
6. Combining two or more trips into a single one.

These methods for reducing car use are almost self-explanatory. The main requirement is that we realize the need for reducing consumption of gasoline. Hence, we should plan ahead before using our cars. We should examine each of the six methods for minimizing use. Not only will gasoline be conserved but we will significantly reduce our gasoline bills.

Most cars on the road today are not in the most efficient operating condition. Spark timing and valve timing may have changed. There may be built-up deposits in the carburetor, altering the air-fuel ratio. Valve seats and piston rings may have become worn. The oil presently in the crankcase may be of the incorrect weight. It is reasonable to expect that the average car engine uses at least 15 percent more gasoline than it should. Some auto dealers and a few service stations do have equipment that can detect improper operation of a car engine and can make the necessary corrections. It is true that it may be rather difficult, particularly in smaller cities and towns, to find either an auto dealer or service station which has the proper equipment to detect inefficient operating conditions. However, if the demand became sufficiently great, it seems certain such equipment would be installed. If all automobile engines were in the proper operating condition, it would be possible to save approximately 4,000,000 gallons of gasoline per day. Assuming that the average driver uses 800 gallons of gasoline per year, it is possible for him or her to reduce the cost of gasoline around 175 dollars per year. This should be more than ample to pay for the inspections and adjustments of the engine.

In addition to keeping the engine in proper operating condition, savings can be made by maintaining the correct air pressure in the tires. Underinflated tires waste a considerable amount of energy. Radial tires give somewhat more gasoline mileage than do regular tires. These tires can make significant gasoline savings when the car is used extensively.

Most motorists can make some reduction in the amount of gasoline which they use by improving their driving habits. In recognition of the fact that high speed operation is wasteful of gasoline, the 55 mile per hour speed limit was established at the time of the gasoline shortages. Although this limit is still in effect, it is ignored by many motorists, who thus waste significant quantities of gasoline. Gasoline is wasted by many urban drivers who drive in a jack rabbit method (i.e., sudden starts and stops). In general, not only for urban but also for highway driving, savings in gasoline can be made by anticipating changing road conditions. It is desirable to slow down when approaching a curve rather than to approach the curve at full speed and then jam on the brakes. It is desirable to slow down when approaching a steep downgrade rather than to approach the downgrade at high speed and to continue at high speed until it becomes necessary to jam on the brakes to prevent excessive speeds.

Some savings in fuel can be made by avoiding excessive idling of the engine. Too many motorists keep their engines idling, for example, while they wait for a freight train to pass through a crossing or while they do a short errand. The automotive air conditioner consumes a significant amount of energy. It can reduce gasoline mileage by as much as 10 percent. Thus, it behooves the motorist wishing to save fuel and also money to minimize the use of the air conditioner.

23.5 CONSERVATION IN RESIDENTIAL AND COMMERCIAL USE

Care must be taken in characterizing the percentage of the total energy supply which is used in the residential and commercial sectors. As may be seen from Table 2-2, approximately almost a quarter of our energy supply goes to the residential and commercial sector. Over a fifth of our energy supply goes to electrical conversion and power line losses. Since about a third of our electrical energy is used by the residential and commercial sectors, the same amount of conversion and line losses occur because of the demands of the residential and commercial sectors. Thus the residential and commercial sector is responsible for almost a third of our total energy uses. Approximately two-thirds of the energy chargeable to these two sectors is chargeable to the residential sector. This means that individual citizens have control in their homes of approximately 20 percent of the national energy supply. This is in addition to some 12 percent of the national energy supply which is used for automotive purposes. Thus the opportunity exists for the individual citizens of this country to have a significant impact on how we use one-third of the nation's energy.

Most homes contain many small electrical devices, such as electric can openers, electric tooth brushes, and electric carving knives. These devices are appreciated for their convenience and are an indication of the extent of the mechanization of our society. The energy consumption by these devices is so small, however, that it is not possible to make any significant savings of energy even if their use were completely discontinued.

Most of the energy used in these two sectors is used for space heating, for air conditioning, for cooking, for refrigeration, for water heating, and for laundry purposes. The factors influencing the amount of energy used for space heating and air conditioning are:

1. Building construction.
2. Efficiences of the heating and air conditioning systems.
3. Operating conditions with the space.

With the growing energy shortages and sharply increasing prices, it is essential that heat transfers to and from a building be minimized. To minimize heat flows it is desirable that the ceiling in the top story rooms of the building contain a minimum of 6 inches of insulation and the sidewalls approximately 4 inches. When the building does not contain a basement, at least 2 inches of insulation should be placed beneath the first floor.* This amount of insulation is easy to install as a building is being constructed. Generally in existing structures, the desired amount of insulation can be installed above the top ceiling and beneath the bottom flooring without too much trouble. In some types of existing structures it may be difficult to install insulation satisfactorily in the sidewalls, particularly if no vapor barrier had been installed. (Without a vapor barrier, vapor may condense and freeze in the insulation.) Compared with normal building wall construction, glass conducts heat rather readily. Hence, glass areas in the walls of a building should be minimized. When glass is used, heat losses can be minimized by the use of thermopanes† or storm windows. Weather stripping should be installed around all windows and doors and all cracks sealed with caulking to minimize flows into or out of the building. It should be noted, however, that when a furnace is used to supply heat, provisions must be made for it to receive sufficient air to ensure good combustion.

Although cost consideration is important in selecting both heating and air conditioning equipment, still more important in the long run is the efficiency of the equipment, together with its anticipated useful life. In general, the lower priced air conditioners require significantly more power to produce a given amount of cooling. Quite frequently, the lower priced furnaces do not have the desired amount of heat transfer surfaces and hence waste energy. Thus, the added cost of energy required for low priced heating and air conditioning equipment more than offset their original lower costs.

Other factors to be considered in heating and air conditioning of buildings is the selection of the correct size of equipment and the design of the distribution system (ducts for an air system, pipes for a steam or hot water system). Competent persons should be selected to design and install the complete system.

* These figures vary with the type of insulation used and also with the region in which the building is located. For very cold regions, much more insulation should be used.

† Two panes of glass tightly sealed in the window frame with a dead air space between the two panes.

Rather extensive tests have shown that the average household oil furnace burns approximately 15 percent more fuel than it should, because of improper adjustment of operating conditions. It should be noted that when too little air is supplied to the furnace, the combustion is incomplete with a significant amount of fuel being wasted. When too much air is used, this excess air is heated and then discharged up the chimney, carrying a significant amount of energy with it. In addition, if the so-called burner does not atomize the oil sufficiently, there will be some incomplete combustion. The average home oil-fired furnace probably burns at least 150 dollars more oil per year than it should because of improper furnace adjustment. The potential savings should be more than sufficient to pay for furnace inspection and adjustments. Although some furnace servicemen are capable of detecting improper operating conditions, many are not. Perhaps if there were a greater demand for this type of service, more furnace mechanics would be trained for this type of work.

Gas is easier to burn than oil, and gas furnaces are more efficient. But even with gas furnaces it is possible to improve the efficiency by making proper adjustments. For both types of furnaces, the heat transfer surfaces must be kept clean.

It should be recognized that significant savings of energy can be made by keeping the temperatures in the building a few degrees below the customary value in the winter and a few degrees above the normal in the summer. Rooms not in use should be closed off from the rest of the building. When the building is not to be used for some time, say even for a day, it should be supplied with only sufficient heat in winter to prevent freezing. It is not true, as some believe, that there is no significant saving when heat is cut off from a building. The statement has been made that savings realized during the time when the thermostat is set low are nullified by the additional amount of heat that is subsequently supplied to warm the building. This is not true. The amount of heat to be supplied to a building must equal the heat lost from the building. During times of low temperature within the building, less than the normal amount of heat is lost. Furthermore, a heat loss from the building greater than normal can not occur unless the temperature within the building is above normal. There is, however, one possible compensating factor. If an attempt is made to warm up the building in a very short period of time by overloading the heating system, there may be a decrease in the efficiency of the heating system. This may tend to partially offset the lower heat loss taking place at low building temperatures.

Heating loads can be reduced in winter by allowing sunlight to enter the building through windows to the extent that is feasible. (It is recognized that there may be some objections since sunlight can cause some fading of objects within the building.) At night in winter, windows should be closed off by blinds or drapes to minimize heat losses from the building. In summer, sunlight should be prevented from entering the building to minimize air conditioning loads.

Provided that the outside air is not dirty, polluted, or of too high humidity, it is desirable to bring in night air in the summer to help reduce the air conditioning load. In the summer, it is desirable to ventilate the attic space unless natural ventilation does this satisfactorily. (Normally in summer, without ventilation the attic temperatures become so high that there will be some heat flow down into the rooms below it in spite of the use of much insulation.)

Many homes are built with fireplaces. A fire in a fireplace gives the cheery, cozy feeling which so many people like. But, as normally operated, fireplaces are very inefficient. Although some heat is radiated into the room, much of the energy of the wood goes up the chimney. When there is a low fire in the fireplace, much of the warm air in the room is drawn up the chimney. When the fireplace is not being used, the chimney damper must be closed to prevent the flow of warm room air up the chimney. As discussed in Chapter 18, devices are being developed to improve fireplace efficiencies, thus presenting an opportunity to use wood to conserve our oil and/or natural gas for home heating.

Space does not permit a detailed discussion here on how energy can be saved in other household and commercial appliances. In general, assuming that it has been decided that the appliance is needed, a study should be made of the energy demands of the appliance. For example, will an automatic washer leave so much water in the clothes that the dryer will use excessive amounts of energy in the drying operation? Much energy can be saved if such equipment as clothes washers and dish washers are not operated until full loads are available. Hot water heating consumes a significant amount of energy. Some of this energy can be saved if the temperature to which the water is heated is lowered and if we are careful in the use of hot water. We have the opportunity to make rather small savings in the use of many of our larger household and commercial appliances. Taken together, the savings can be significant.

Recognizing the potential for savings of significant amounts of energy in house heating and cooling, some agencies of the government are subsidizing, by direct grants or by tax credits, efforts to minimize heat losses and gains from homes. In addition, there are many publications available to help the home owner to conserve energy. If every home owner and every home occupier would recognize the necessity for energy conservation and would make real efforts to do so in the various activities within the home, there would be a very significant reduction in the amount of energy used in this country.

23.6 SUMMARY ON ENERGY CONSERVATION

In this country we are demanding ever-increasing amounts of energy. Part of this increase in demand is caused by our growing population. But most of the increase is attributable to the increasing per capita use of energy. Part of the growth in the per capita demand for energy results from the increasing use of energy-intensive machinery and other devices to do our work for us. The rest of the growth in energy demand results from wasteful and careless use.

Thus the opportunity exists for us to make significant reductions in our use of energy. To do this we should

1. Examine the overall situation very carefully before expanding on the use of energy consuming devices.
2. Examine very carefully our uses of energy to make certain that the usage is truly necessary and then we should make certain that we do use energy efficiently.

With the exception of a small portion of fossil fuels which is used in the production of plastics and other chemicals, our energy supply is used in four sectors. One of these sectors, electric power, is using a rapidly growing portion of total energy supply. The energy produced in this sector, after the inherent generation and transmission losses are deducted, is delivered to the other three sectors, the industrial sector, the transportation sector, and the residential and commercial sector.

In all four sectors the use of energy can be minimized by the following measures:

1. In selecting an energy-using device, a careful examination should be made of all possible devices in order that one can be chosen which will use the minimum amount of energy consistent with performing its function in an acceptable manner.
2. Consideration should be given to a choice of the type of energy to be supplied to the device to make certain that the type of energy is best suited to the device and should remain so for the life of the device.
3. The device should be kept in first class operating condition at all times.
4. The operators should be capable and willing to operate the device in the most efficient manner.

On the one hand, it is possible to decrease the amount of energy which we use in this country by shifting work loads from machines and then making certain that when we do use energy, we use it wisely and well. On the other hand, if we increasingly shift work loads to machines and other devices, if we do not choose these machines and devices well and do not operate them in a most efficient manner, our demands for energy will become so great that economic chaos may result. To avoid the dislocations which will arise with either of these extremes, we should expand our use of energy-consuming devices only when such an expansion is fully necessary. Above all, we should scrutinize all of our uses of energy to make certain that we are using energy wisely and well.

SUPPLEMENT A

ILLUSTRATIVE EXAMPLES

Example 23-1. A car owner, driving 15,000 miles per year, averages 16 miles per gallon. By proper engine maintenance and careful driving, 20 percent of the fuel can be saved. If gasoline costs $1.40 per gallon, determine the yearly savings.

Solution. The amount of gasoline used per year equals

$$\frac{15{,}000}{16} = 937.5 \text{ gallons}$$

The potential savings equals

$$937.5 \times 0.20 = 187.5 \text{ gallons}$$

The dollar savings equals

$$187.5 \times 1.40 = \$262.50 \qquad \text{answer}$$

Note: The dollar savings should be more than ample to keep the car in first class operating condition.

Example 23-2. An air conditioner is added to the car in Example 23-1 which reduces the gasoline mileage by 10 percent. The air conditioner is used for 4,000 miles travel per year. Determine the annual cost of gasoline chargeable to the air conditioner, assuming the original condition of the car.

Solution. The amount of gasoline required for the car without air conditioning equals

$$\frac{4{,}000}{16} = 250 \text{ gallons}$$

The gasoline required for air conditioner equals

$$250 \times 0.1 = 25 \text{ gallons per year}$$

The cost of this extra gasoline equals

$$25 \times 1.40 = \$35.00 \qquad \text{answer}$$

Example 23-3. Assume that there are 40,000,000 cars similar to that in Example 23-2. Determine the number of barrels of gasoline required per year by the air conditioners on these cars.

Solution. The gasoline used per year equals

$$40{,}000{,}000 \times 25 = 1{,}000{,}000{,}000 \text{ gallons}$$

or

$$\frac{1{,}000{,}000{,}000}{42} = 23{,}800{,}000 \text{ barrels} \qquad \text{answer}$$

Example 23-4. A 150 unit apartment building is heated by an oil furnace. During the heating season, each unit requires, on the average, 20,000 Btu per hour. Assume the heating season to be 175 days in length. The oil has a heating value of 140,000 Btu per gallon and costs $1.30 per gallon. By proper maintenance, the efficiency of the heating system can be increased from 60 percent to 72 percent. Determine the savings in fuel, dollars per year.

Solution. The heat presently supplied per gallon of oil is

$$140{,}000 \times 0.6 = 84{,}000 \text{ Btu}$$

The heat required per year per apartment equals

$$20{,}000 \times 24 \times 175 = 84{,}000{,}000 \text{ Btu}$$

The amount of oil required equals

$$\frac{84{,}000{,}000}{84{,}000} = 1{,}000 \text{ gallons}$$

For the 72 percent efficiency, the oil required per apartment is

$$\frac{84{,}000{,}000}{140{,}000 \times 0.72} = 833 \text{ gallons per year}$$

The savings in oil per apartment equals

$$1{,}000 - 833 = 167 \text{ gallons}$$

The savings for the whole building equals

$$167 \times 150 = 250{,}050 \text{ gallons}$$

The cost of this oil equals

$$25{,}050 \times 1.30 = \$32{,}570 \qquad \text{answer}$$

Example 23-5. By careful operation, the efficiency of a power plant is increased from 38 percent to 39 percent. The average load on the power plant is 1,500,000 kW. The power plant burns coal costing 38 dollars per ton and having a heating value of 12,800 Btu per pound. Calculate the yearly savings made in the cost of coal.

Solution. The present plant output per pound of coal is

$$12{,}800 \times 0.38 = 4{,}864 \text{ Btu}$$

Per kW output the coal required per year equals

$$\frac{8{,}760^{*} \times 3{,}413}{4{,}864} = 6{,}147 \text{ lb.}$$

* 8,760 hours per year.

For an efficiency of 39 percent, the coal required per year per kW output equals

$$\frac{8{,}760\times 3{,}413}{12{,}800\times 0.39}=5{,}989 \text{ pounds per year}$$

The savings in coal per year per kW output equals

$$6{,}147-5{,}989=158 \text{ pounds.}$$

For the average output of 1,500,000 kW the savings in coal is

$$1{,}500{,}000\times 158=237{,}000{,}000 \text{ pounds, or } 118{,}500 \text{ tons}$$

The cost of this coal is

$$118{,}500\times 38=\$4{,}503{,}000 \qquad \text{answer}$$

SUPPLEMENT B

PROBLEMS

23-1. Approximately 10 percent of the energy used by the residential-commercial sector is for water heating. Assuming that a concerted conservation effort will result in a savings of 15 percent of this energy, calculate the savings to be made by such an effort, expressed in equivalent millions of barrels of oil per day. Note: See tables in Chapter 2.

23-2. For the conditions of Problem 23-1, express the savings as a percentage of the total national energy use.

23-3. A truck operator drives 55,000 miles per year. He averages 8 miles per gallon. By proper engine maintenance and careful driving, he can make a reduction in his fuel use of 15 percent. Assuming that the fuel costs $1.40 per gallon, calculate his yearly fuel savings.

23-4. A commercial establishment has, on the average, 180 lights in use eight hours per day. The average wattage is 150. Calculate the monthly savings which can be made by substituting 100 watt bulbs. Assume that electrical energy costs six cents per kilowatt hour.

23-5. Calculate the reduction in air conditioning load, Btu per hour, for Problem 23-4.

23-6. Do you recommend that the larger bulbs of Problem 23-4 be used in winter since they will help to meet the heating load? Why?

23-7. To meet the peak load conditions, a power company uses old generating equipment which has a heat rate (Btu supplied per kWh) of 11,500 Btu per kWh. The average load to be carried by this older equipment when operating is 1,200,000 kW. If newer equipment were to be installed for peak load operation, it would have a heat rate of 8,800 Btu per kWh. In both cases, the fuel to be used is coal, having a heating value of 13,200 Btu per pound. Calculate the average number of tons of coal to be saved per hour by using the newer, more efficient generating equipment.

23-8. Would you recommend continuing the use of the older generating equipment in Problem 23-7? Why?

23-9. Can the use of smaller, lighter weight cars by the general public be greatly increased? If so, how?

23-10. Is it possible that the general public will so operate their cars that there will be a significant reduction in the consumption of gasoline? If so, how can this be accomplished?

24

Energy and the Environment

24.1 INTRODUCTION

Substantially all of the energy we presently use comes from fossil fuels, nuclear fuel, and from hydropower. It is impossible to procure the fuels from nature without disturbing the earth in which they are located. Hydropower, in general, requires the damming of rivers, thus flooding a considerable amount of land. Many of the proposed alternative energy sources, particularly geothermal, tidal power and shale oil, can cause very serious environmental problems as they are being developed. Different kinds of environmental problems may arise in the development of solar energy, wind energy, and ocean thermal gradient energy.

Additional environmental problems may occur as raw energy material is transformed into readily usable energy. Furthermore, in many instances, we can cause environmental problems as we use the energy. Over the years, many very serious environmental dislocations have been caused by our demand for energy. There are four approaches which may be taken concerning the energy-environmental conflict:

1. We can greatly reduce our demand for energy, thus minimizing the potential for environmental disturbances.
2. We can disregard the serious, and, in many cases, irrevocable damage which can occur if we take energy from nature without using any precautions to protect the environment.
3. We can try to hold the environmental damage to a minimum and attempt to repair it as it occurs. It must be recognized, however, that this course of action will be a costly one.
4. We can choose energy sources which are least polluting in procurement and use. It must be recognized, however, that generally such sources are very expensive and may demand a sizeable portion of our resources for their use.

In previous chapters, environmental problems were discussed as the various energy sources were studied. It is the objective of this chapter to review and contrast these problems. This will be done by identifying the problems associated with the following three categories:

1. Energy Procurement.
2. Energy Transformations.
3. Energy Use.

It is not possible to present here in detail all of the environmental problems associated with all of the possible energy sources, but the broad problems can be discussed.

24.2 ENERGY PROCUREMENT

1. Coal. Although coal is, by far, the most abundant of our fossil fuels, many environmental problems are associated with mining it. Approximately half of our coal presently is obtained by strip mining. The overburden is first removed from above the coal and placed in piles. In mountainous regions it is very difficult to prevent rain from washing this overburden down onto the land below it. In flatter regions, after the mining operations have been completed, it is possible to restore the land to satisfactory conditions and to revegetate it, provided sufficient water is available. To completely restore the land may be very expensive and will be reflected in the cost of the coal.

Underground mining is not only expensive and dangerous but may result in cave-ins of the land above the mines. In addition, water pumped from the mines and discharged into streams and rivers may be quite acid. Before shipping the coal, much of the rock and slate mined with the coal is separated from it. The disposal of this refuse is difficult without causing adverse environmental effects. It is also difficult to suppress the dust existing around the mine.

2. Oil and Natural Gas. Other than objections to the appearance of oil rigs dotting the landscape, oil and gas production does not, in general, cause extensive adverse environmental problems. Years ago in this country, and even now in some parts of the world, gas originating in the production of oil was flared off (i.e., burned), causing a considerable amount of pollution in the vicinity of the oil well. This practice was abandoned in this country quite some time ago as a market was found for the gas. There have been a few oil well and gas well explosions. When these occurred on land, fires resulting from the explosion did cause a considerable amount of pollution. However, these fires normally were extinguished in a relatively short period of time. There have been some explosions and some leaks in off-shore wells. In a few cases, such as the one near Santa Barbara, California, the results have been disastrous as oil slick covered the beaches for many miles. Because of this potential difficulty, various environmental groups are trying to block further off-shore drilling, particularly off-shore of the eastern United States, where large amount of oil are believed to exist.

3. **Uranium.** Even though a ton of uranium ore may contain only about five pounds of uranium, because of the high energy content of the uranium, very much less uranium ore has to be mined compared, for example, with coal having the same energy content. However, as the demand for uranium grows, the disposal of the rock associated with the uranium may become somewhat of a problem. Much of the uranium is found in desolate regions and hence the rock disposal problem is not presently a critical one.

4. **Hydropower.** Except for the very few run-of-the-river hydraulic plants, hydropower plants require large reservoirs to produce the desired head and to provide water storage for times of low river flow. This has meant the flooding of valuable agricultural lands in some cases and scenic areas in other cases. Frequently it has been necessary to displace many residents located at the reservoir site. Although they seldom occur, there have been some dam failures, resulting in destruction of life and property downstream.

5. **Geothermal Energy.** The magnitude of the pollution problems associated with geothermal energy depends on the nature of the geothermal source (i.e., is it steam or water, does it contain many dissolved gases and, for water, what is its salinity). The geothermal steam at the Geysers contain many gases, the most troublesome one being hydrogen sulfide, which is difficult to remove. The salinity of some geothermal brines is very high, in some cases being about six times that of ocean water. This not only creates very serious problems in the heat transfer equipment of the power generating system but also creates a difficult water disposal problem. In addition, when vast quantities of geothermal water are withdrawn from the earth there is a possibility of ground subsidence. It has been suggested that the withdrawal of large quantities of geothermal water could trigger earthquakes in unstable regions. These effects can be minimized if the water is reinjected at some distance from its point of removal.

Another environmental problem associated with geothermal power production is that of thermal pollution.* Since the operating temperature range in geothermal power plants is low, the thermal efficiency is correspondingly low. Thus much more heat must be rejected by geothermal power plants than by fossil fuel power plants having the same output.

6. **Shale Oil.** Although shale oil has the potential of supplying much more oil than our known petroleum reserves, the environmental problems associated with shale oil production are so large that it is doubtful if significant amounts of shale oil will be produced in the next decade or two. When the oil shale is strip-mined, the opportunity exists for returning the spent shale to the mined-out holes. The overburden, which was removed initially, can be replaced on top of the spent shale and revegetation then attempted in order to restore the land to acceptable conditions. Revegetation may be quite difficult

* Thermal pollution normally refers to the vast quantities of circulating water discharged from the condensers at sufficiently high temperatures to cause serious environmental problems when dumped indiscriminately.

since most of the oil shale exists in areas of very low rainfall. Most of the water which is available is now committed to other uses, making it very difficult to obtain the vast quantities necessary for large scale revegetation operations.

Present indications are that much of the mining of oil shale will be underground. Since underground disposal of the spent shale is not feasible, its disposal in an acceptable fashion is a major problem. Although most of the oil shale is located in rugged terrain, which offers valleys for possible spent shale disposal, it is very difficult to fully contain the spent shale once it is dumped. The difficulty of revegetation on top of the spent shale from underground mines is the same as that for the strip-mine operation, largely the lack of water.

The retorting of the shale to produce oil is accompanied by problems of noise and dirt in the ore crushing process and discharges from the retorting operation itself. Here again, the obtaining of the necessary amount of water for the retorting operations is a serious problem.

7. Solar Energy. When solar energy is collected for space heating or for hot water heating, the collectors generally cover part of the building to which heat is to be delivered. As such, they do not create environmental problems. Although not an environmental problem in the strict sense, a serious legal problem may develop relative to the possible shading of collectors by neighboring trees and buildings, thus preventing them from receiving the maximum amount of solar energy. Because of their very low efficiency, solar cells must be spread over very large areas to produce large amounts of electric power. However, it does not appear that solar cells will be used for large-scale production of power, certainly not in the near future. Hence, the problem of diverting land area for this purpose will be a very minor one.

There is a possibility that focused sunlight may be used to generate steam, directly or indirectly, for electrical energy production. The most promising region of this country for such an operation is the desert southwest. Here, the land required for solar energy collection is substantially worthless and hence its use for this purpose would not detract from other possible usages. The proponents of this method of using solar energy feel that this use would not have detrimental environmental effects since the amount of energy transformed into electrical energy and delivered from the region approximately equals that portion of the solar energy which is normally reflected. Even when cooling towers are used, large quantities of water would be required for the condensers of the proposed power plants. Because so little water is available in this region, it may be necessary to pump water from the Gulf of California. The safe disposal of the salts in this ocean water will be a problem. Assuming that cooling towers are used, the vapor which they put into the air should be of some benefit to the region.

8. Wind Power. It does not appear that the use of wind power will create serious environmental problems although windmill noise may be objectionable. There probably will be some objections raised about the appearance of many wind power plants several hundred feet high. There is some feeling that the use of a very large number of wind power plants in a restricted region could, by extracting large amounts of energy from the wind, have a significant effect on the weather. Because of the large number of variables involved, no models have been made which can demonstrate whether or not this is true.

9. **Marine Energy.** With the exception of those very limited areas where the difference in elevation between high and low tide is great (i.e., at least 15 to 20 feet) tidal power does not appear to be economically feasible. Where tidal power is feasible the one environmental problem which arises is the vast amount of land needed for storage basins.

Present indications are that neither ocean currents nor wave power use is economically feasible for the continental United States and hence their possible adverse environmental effects do not need to be considered here.

There is a possibility that there may be some use of the thermal gradients in the oceans in the next two or three decades. In general, it does not appear that environmental problems will be caused by using these thermal gradients. Some suggestions have been made, however, that if a large number of such plants were to be operated in a warm ocean current, such as the Gulf Stream, the removal of energy from the stream could produce some changes in the stream. Information is lacking to evaluate the effects of these changes.

The use for power production of the difference in the salinity of fresh water at the mouths of rivers and that of the oceans into which they discharge, has not been carried much beyond the discussion stage. Hence, it is not certain that it will ever become economically feasible. If this approach becomes feasible, it appears that a serious environmental problem would be created. A dam would have to be built at the river mouth, thus interfering with both free river flow and with river-ocean ship traffic.

24.3 ENERGY TRANSFORMATIONS

Strictly speaking, an oil refinery does not transform energy from one form into another. But it does convert the crude oil into products that can be used for specific purposes, such as gasoline and diesel fuels for transportation, heating oils, and boiler fuels. Oil refineries have been notorious for the pollution which they have caused in the form of odors, toxic gases, and particulates. Oil refineries require large quantities of water for their operation. The spent water, in the past, has caused some pollution as it was discharged without proper treatment. At present, however, most of the oil companies have initiated extensive programs to control all types of pollution in their refineries. The cost of these anti-pollution efforts has been significant and has added to the cost of petroleum products.

Associated with refinery operation is the transportation of the oil from the oil fields to the refinery, particularly in tanker ships. Serious oil spills have occurred in transferring the oil to and from the ship and on the seas.

There are several environmental problems associated with power production using either fossil or nuclear fuels. One of these is power plant siting. To minimize transmission line losses, it is desirable to locate the power plant near the center of the load. When the load is a concentrated one, this means that it is desirable to locate the plant in a built-up area. The power plant requires a considerable amount of land—land which may have the potential for several other usages. Then too, the residents of the area may object to a power plant, particularly if it is a nuclear one, being built in their neighborhood. To avoid these objections, power plants generally are built in less congested areas. Here the required land has less potential for other uses. And there are few neighbors to object to the construction in their vicinity. But there is a growing shortage of rural land suitable for power plant construction. Particularly when coal is the fuel, rail connections must be available to bring the large quantities of coal which will be burned. The plant must be located so that large quantities of water will be available for the condensers. It has been estimated that the total electric generating capacity in this country in the year 2000 will be about three times the present capacity. This means that power plant siting will become an ever-increasing problem.

If coal is burned for power generation in the increasing amounts predicted, the problems of atmospheric pollution will be very severe unless extensive and expensive efforts are undertaken to minimize the pollution. In particular, it is essential that devices be incorporated in coal burning power plants to minimize emissions of particulate matter, sulfur dioxide and oxides of nitrogen. These problems are minor when oil is the fuel and are even smaller when natural gas is burned in steam power plants.

Another environmental concern resulting from the burning of fossil fuels is the "greenhouse effect." The burning of fossil fuels both in power plants, furnaces of all kinds, and internal combustion engines, produces carbon dioxide. Although plant life uses much of this carbon dioxide in their growth, some scientists state that there is an alarming increase in the amount of carbon dioxide in the atmosphere. Since carbon dioxide absorbs and then radiates back much of the natural radiation from the earth, it is predicted that, as we burn an increasing amount of fossil fuels, there will be an increase in the surface temperature of the earth. There is some prediction that, ultimately, there will be a sufficiently high increase in the surface temperature to cause melting of the polar ice caps. This, say some, will cause a sufficient increase in the water level in the oceans to flood all the coastal cities of the world. In contrast to this prediction, there are those who feel that excessive amounts of carbon dioxide in the earth's atmosphere will absorb and reflect back much of the solar energy normally striking the earth. If this is true, in spite of the greenhouse effect, the temperature of the earth's surface will gradually decrease. It is not possible at this time to predict the true net effect of an increase in the amount of carbon dioxide in the atmosphere.

A very major environmental problem associated with steam power generation is that of thermal pollution. A power plant burning 500 tons of coal per hour will reject to the circulating water in its condensers every hour energy equivalent to over 250 tons of coal. In turn, the water must dispose of this energy. If the circulating water were to be returned to the river or lake from which it was drawn, it would produce an excessive temperature rise unless the body of water was very large. Hence it is desirable to cool the water and then reuse it in the condensers. When sufficient land is readily available for the construction of a cooling lake, this is the least expensive method of cooling the circulating water. Since the cooling of the water is accomplished largely by evaporation, there may be a significant increase in the humidity of the air adjacent to the lake, particularly during periods of stagnant air conditions. For large power plants, the cooling lakes may be several thousand acres in size. The Environmental Protection Agency has objected to diverting so much land for this use and has, very strongly, urged the use of cooling towers. In the normal type of cooling tower, the wet cooling tower, much of the cooling is accomplished by evaporation. This evaporation of water will tend to increase the humidity of the air similar to that for the cooling lakes. Cooling towers are much more expensive, initially, than cooling lakes, except in built-up areas where the land required for the lakes may be very expensive. Furthermore, cooling towers are expensive to operate. Thus, the cost of generating power is higher when cooling towers are used.

For very large power plants, the evaporation loss from either the cooling lakes or cooling towers may be as high as two to three million gallons of water per hour. Particularly in the drier regions of the country, it may be difficult to obtain this amount of make-up water. As more and more steam power plants are built, the dual problems of obtaining sufficient water, even for evaporative cooling, and the increased humidity of the air caused by the evaporation will be intensified. These problems can be minimized by use of dry cooling towers, in which the circulating water is cooled by blowing air over heat transfer surfaces. Although the air in the immediate vicinity of the power plant will be heated significantly, this will not be as objectionable as the humidity increase caused by evaporative cooling. However, dry cooling towers are much more expensive initially than wet cooling towers and are more expensive to operate. Furthermore, because dry cooling towers normally do not cool the water to as low temperatures as do cooling lakes and wet cooling towers, the steam pressure in the condensers is higher, with the consequence that ten to fifteen percent more fuel must be burned in the plant for a given electrical output.

Another possible environmental problem associated with coal burning power plants is that of disposal of the ash after combusion. In very large plants, there may be 25 to 50 or even more tons of ash to be disposed of per hour. Much of this ash from plants using pulverized coal is in the form of fly ash. In certain areas it has been used to make light weight masonry blocks. In rural areas, it is frequently being used as a landfill. In other areas, its disposal without causing adverse environmental effects is costly.

Present day nuclear power plants have a significantly lower thermal efficiency than do fossil fuel power plants. Hence the nuclear plants must reject perhaps 40 percent more heat, thus greatly aggravating the thermal pollution problem. Because of all the safety precautions which have been built into the various processes, such as ore refining, uranium enrichment, fuel element manufacturing, nuclear plant operation, and fuel reprocessing, the adverse environmental effects caused by radiation from the various processes is normally within acceptable limits. It is theoretically possible, however, in spite of the use of many safeguards, that a serious accident could occur in a nuclear power plant which would release radioactive material. The possibility of such an occurrence is almost nil, but perhaps additional safeguards should be used to lower still further the possibility of such an occurrence. There is a very low possibility of nuclear plant sabotage and of theft of radioactive material. There is also the environmental problem of the disposal of radioactive waste. Although many safeguards are now being used, each of these possible problems should be studied carefully to see how additional safeguards can be used effectively.

As discussed in Chapter 20, it seems evident that there will be increasing efforts made to produce gas and oil from coal. Unless proper precautions are taken in such processes, obnoxious gases as well as particulate matter will be discharged into the atmosphere and the waste water will contain polluting compounds. The solid waste from these processes may also contain polluting material.

When solid waste, particularly municipal refuse, is used for its energy content, care must be taken to avoid adverse environmental effects. These include odors, particulate discharges, and noxious gases which can result from either pyrolysis or incineration of the waste.

24.4 ENERGY USE

In general, energy is used primarily either in the form of electricity or as a fuel for heating or power producing. Electrical energy, in itself, does not produce environmental problems as it is being used. It is true that adverse environmental effects may be caused in the generation of the electrical energy. But these effects can be controlled and minimized much more readily in the power plants than they could be if the same fuel were to be burned in homes and commercial installations.

Formerly coal was used extensively for residential and commercial heating. Unless the coal was anthracite, this use of coal caused excessive pollution. When oil and natural gas became available in large amounts and at low cost, the use of coal for heating was largely discontinued except in large installations where facilities should be available to control pollution.

When the fuel burning equipment is not adjusted properly, the burning of oil can cause significant amounts of pollution. Some fuel oils are high in sulfur. When burned, the sulfur is discharged as sulfur dioxide, a serious pollutant. Generally the burning of natural gas does not cause a serious pollution problem.

A very large percentage of our oil is burned in internal combustion engines. Although the exhaust from diesel engines may, at times, be smoky and have an odor, generally this exhaust does not contain as much of the pollutants such as hydrocarbons, nitrogen oxides, and carbon monoxide as is found in the exhaust from gasoline engines. However, if the diesel fuel is high in sulfur, excessive amounts of sulfur dioxide will be produced.

The gasoline engine, whether used for automotive, truck, bus, or airplane service, is a notorious polluter. Because of the tremendous number of vehicles used in metropolitan areas, the air in these areas has become highly polluted. The Environmental Protection Agency as well as some states have established progressively stricter standards for limiting harmful emissions from gasoline engines. By changes in design and by the use of such devices as catalytic converters, automotive manufacturers have been able to reduce the pollution caused by their engines. But these efforts have added significantly to the cost of automobiles. There is some indication that the addition of pollution control devices definitely reduces the gasoline mileage.

There has been considerable discussion about and some research into using other propulsive devices in place of the gasoline engine to minimize the pollution problem. These proposed devices include the gas turbine and the steam engine. We now possess the requisite technology to use these devices. However, the initial cost of both the gas turbine and the steam engine is much higher than that of the internal combustion engine. There is considerable doubt whether the costs of these units can be reduced sufficiently to make them competitive with the gasoline engine. In addition, the steam engine does not possess the flexibility now demanded by the motoring public. Thus, in spite of the pollution which it produces, it appears that the gasoline engine will continue to be a main source of power for automotive service for many years to come.

Even if the gasoline engine can be designed to minimize pollution, an excessive amount of pollution will be produced unless it is kept in proper operating condition. The automobile operator bears a responsibility along with the engine designer to keep emissions from automotive engines down to acceptable standards.

24.5 ACID RAIN

There is a growing concern about acid rain and its effects on the environment. With the burning of increasing amounts of fossil fuels in steam power plants and in internal combustion engines, there is an increase in the pollutants discharged into the atmosphere. These include sulfur compounds, nitrogen oxides, carbon monoxide, hydrocarbons and particulate matter. Because of the greatly increasing use of fossil fuels, there is an increase in the amount of pollutants discharged in spite of efforts being made to remove them from the products of combustion.

There is a tendency for these pollutants to combine with the moisture in the air to form acids. Thus the rain in the given locality is known as acid rain. As such it is harmful to buildings and, in some cases, to vegetation. Many lakes have become so acid that fish can no longer live in them. This problem of acid lakes is a particularly real one in the northeastern part of this country, especially in upper New York state, and also in the Scandinavian countries. Indications are that acid rain may fall hundreds of miles away from where the pollutants were discharged into the air.

Much is to be learned about acid rain and about the increase in acidity of lakes. At some rain stations, there have been considerable variations in the acidity of the rain from one storm to the next. There is also considerable variation, in some cases, in the acidity of lakes in a given area.

Part of the variation in the acidity of the rain in a given locality can be attributed to variations in wind patterns. In addition, the output of pollutants at their source may vary with time. It should be noted that there are sources of pollutants other than plants and internal combustion engines. Such operations as smelting and coal gasification can produce pollutants. A considerable number of pollutants may come from nature, including volcanoes as well as decaying vegetation.

Apparently there are several factors other than acid rain which contribute to the acidity of lakes. The air may contain pollutants which have not reacted with moisture. As the air sweeps over a lake, some of these pollutants react with the water in the lake. Rivers and streams feeding a lake may bring pollutants which they absorbed from vegetation and from the soil. There is evidence to show that the lake itself can pick up some acidity from the earth and rocks in the lake and from growths in the lake, particularly from bogs.

In summary, there are many factors which contribute to the acidity of air and of lakes. Only part of this acidity can be charged to the burning of fossil fuel. However, because of the adverse effects of the pollutants caused by burning fossil fuel, even more effort should be made to minimize the amounts of these pollutants.

24.6 SUMMARY ON ENERGY AND THE ENVIRONMENT

As we draw energy from nature, we create disturbances in the environment. The adverse environmental effects are at a minimum when we use solar energy and wind power. But energy drawn from these two sources presently is too expensive to use. Furthermore, it is questionable whether their costs can be reduced sufficiently to permit extensive use for many years to come, certainly not in the next decade or two.

In general, we cause serious environmental disturbances as we procure energy from the earth, as we transform the energy bearing material into a usable form and, in many cases, as we use the energy. Whether we like it or not, we, the users of energy, must take the ultimate responsibility for adverse environmental effects associated with energy. Energy bearing materials are taken from the earth solely because we demand energy. The adverse effects on the environment can be minimized only if we will pay the costs. Pollution arising from energy transformations, such as those occurring in refineries and power plants, can be minimized if we, the public, will pay the bill. Furthermore, we are too often careless in our use of energy. As we burn gasoline in our cars and as we burn oil to heat our homes, we pay little attention to avoiding pollution.

The magnitude of the adverse environmental effects associated with energy is, then, dependent on how much energy we demand and on how much we will pay to minimize these effects.

SUPPLEMENT A

ILLUSTRATIVE EXAMPLES

Example 24-1. Coal is produced from a strip mine at the rate of 4,500,000 tons per year. The average thickness of the coal seam is 12 feet. Take the density of the coal as 82 pounds per cubic foot. Determine the surface area mined per year in acres.

Solution. The volume of the coal mined per year equals

$$\frac{4{,}500{,}000 \times 2{,}000}{82} = 109{,}700{,}000 \text{ cubic feet}$$

The area equals

$$\frac{109{,}700{,}000}{12} = 9{,}140{,}000 \text{ square feet, or } 210 \text{ acres} \qquad \text{answer}$$

Example 24-2. The coal in Example 24-1 is 11.2 percent ash. The ash has a density of 42 cubic feet. Determine the volume of the ash to be disposed of per year.

Solution. The tons of ash to be disposed of per year equals

$$4{,}500{,}000 \times 0.112 = 504{,}000 \text{ tons.}$$

The volume of the ash equals

$$\frac{504{,}000 \times 2{,}000}{42} = 24{,}000{,}000 \text{ cubic feet} \qquad \text{answer}$$

Example 24-3. The coal in Example 24-1 is 3.2 percent sulfur. Determine the total amount of sulfur present in the coal.

Solution. The amount of sulfur equals

$$4{,}500{,}000 \times 0.032 = 144{,}000 \text{ tons per year} \qquad \text{answer}$$

Note: This amount of sulfur is approximately two percent of all sulfur produced annually in this country.

Example 24-4. A nuclear power plant, when producing 1,800,000 kW, has an efficiency of 32 percent. It draws circulating water for its condensers from a river at the rate of 1,800,000,000 gallons per day. The normal temperature of the river in the summer is 85°F. Determine the water temperature leaving the condensers.

Solution. Since there are 8.34 pounds per gallon of water, the water flow is

$$\frac{1{,}800{,}000{,}000 \times 8.34}{24} = 625{,}500{,}000 \text{ pounds per hour}$$

Since the efficiency $= W/Q_s$, the heat supplied to the plant per hour equals

$$\frac{1{,}800{,}000 \times 3{,}413}{0.32} = 19{,}200{,}000{,}000 \text{ Btu}$$

Neglecting extraneous losses, the heat rejected in the condenser equals the heat supplied minus the work, or

$$19{,}200{,}000{,}000 - 1{,}800{,}000 \times 3{,}413 = 13{,}060{,}000{,}000 \text{ Btu per hour}$$

Since the specific heat of water is unity, 1 Btu will produce a temperature rise of 1 pound of water of 1°F. Then the total temperature rise equals

$$\frac{13{,}060{,}000{,}000}{625{,}500{,}000} = 20.9°\text{F}$$

$$\text{The final water temperature equals } 85 + 20.9 = 105.9°\text{F} \qquad \text{answer}$$

Example 24-5. A power plant burns 550 tons of coal per hour. The ash content of the coal is 10.8 percent. The gas cleaning equipment has an efficiency of 99 percent in removing ash from the gases leaving the furnace. Assume that 78 percent of the ash appears in the gases leaving the furnace. Determine the number of tons of ash discharged into the atmosphere per day.

Solution. There are 0.108 tons of ash per ton of coal. The ash not removed per ton of coal equals

$$0.108 \times 0.78(1.00 - 0.99) = 0.00084 \text{ tons.}$$

The tons of ash discharged per day equals

$$0.00084 \times 550 \times 24 = 11.12 \text{ tons per day} \qquad \text{answer}$$

SUPPLEMENT B

DISCUSSION QUESTIONS

24-1. Make a list of all possible adverse environmental effects caused by mining coal, transporting it to a power plant, producing electric energy and delivering this energy to our homes.

24-2. Make a list of all possible adverse environmental effects caused by producing oil, transporting it, refining it, and using it on

   a. Heating homes.
   b. Automotive propulsion.

24-3. Same as Problem 24-1 except use uranium.

24-4. Discuss the desirability of increased governmental regulations to minimize automotive engine pollution relative to both the design and use of the car.

24-5. Pollution can be reduced significantly by reducing energy consumption. Discuss the desirability of rationing all forms of energy to accomplish this purpose.

24-6. If strict pollution control is to be used to minimize adverse environmental effects, how can the average citizen be made to realize that he must pay for these controls?

24-7. Is it desirable that the extent of the use of pollution control measures be decided by a vote of the people?

25

General Summary

25.1 AN OVERVIEW

As we have learned to obtain energy from nature, we have been able to make very drastic changes in our standard of living. We have shifted work loads in all walks of life from the backs of mankind to machines. We have provided material comforts for ourselves undreamed of even a few decades ago. We have been blessed in this country with large supplies of coal, oil, natural gas, and with many suitable sites for hydropower development. Furthermore, energy from these sources has been so cheap that we could use vast quantities without seriously upsetting our personal budgets. Because of this a large segment of our population has come to feel that almost unlimited amounts of low priced energy are its birthright. It is difficult for the general populace to realize that we are facing an ever-growing shortage of energy and that the energy which is available to us will be increasingly expensive. A review of the material presented in this book should make several factors almost self-evident:

1. We are demanding ever-increasing amounts of energy, not only because of our population growth, but also because our per capita demands are increasing significantly.
2. Our domestic supplies of oil and natural gas, which furnish 72 percent of our energy needs, are dwindling. It has been predicted that by the end of this century we will have consumed most of our known domestic oil and gas resources.
3. We are importing oil and, to a limited extent, natural gas in ever-increasing quantities. There is some indication that within the next two to three decades the cost of imported oil will average over four hundred dollars for every man, woman, and child in this country.
4. Although our domestic supplies of coal are plentiful, major problems involved in mining the coal and in burning it must be solved.

5. Particularly if the fast breeder reactor can be fully developed, nuclear fission offers the potential for supplying a large portion of our energy needs. However, several questions relative to nuclear power production safety have not been answered to the satisfaction of many citizens. Hence, it is uncertain how fast and to what extent nuclear fission power can alleviate the future energy shortage.
6. It does not appear that all of the other potential energy sources combined will make a large contribution to our energy supply in the next decade or two.
7. It does not appear to most observers that fusion power, in spite of its tremendous potential, can make a significant contribution to our energy supply before the end of this century, if even then.

In view of the above, we should raise the question: The energy situation, is it a crisis, a dilemma, or just another problem?

25.2 A CRISIS?

Some efforts have been made toward energy conservation but, in general, they are rather feeble. It is true that some industries, perhaps lured by the opportunity to reduce production costs, have made significant attempts to reduce their energy consumption. Some home owners and some builders have endeavored to save energy by using more insulation. It is also true that the total energy consumption in this country dipped after the Arab oil embargo and during the recession which followed it. Although part of this drop was due to conservation efforts, most of this drop was caused at first by shortages of oil and then, almost wholly, by the recession itself. Following the economic recovery, energy consumption resumed its upward spiral. The 55 mile per hour speed limit, established to conserve gasoline, is ignored by most motorists. Although there has been a loud outcry against the increased cost of electrical energy, its use has continued to increase.

Our production of oil is decreasing. To meet our increasing demands, we have increased our oil imports greatly. Even by sale of our manufactured goods abroad, particularly armaments and farm products, we have still been unable to pay for this imported oil. How much longer can we continue to do this?

It seems as if we are drifting along, hoping that something will happen to give us the energy we feel we need. We are making some efforts to conserve energy. With the support of governmental agencies, such as DOE, we are making some efforts to develop alternative sources. But the combined present efforts are not sufficient to greatly affect our energy situation in the near future. Thus, we appear to be drifting from a bad situation to a worse one. Energy prices will continue to increase. We will rely more and more on imported oil. When we can no longer pay for the imported oil or when an embargo is slapped on our foreign oil, we will face a real crisis. Factories will be closed.

The unrestricted use of cars will be curtailed. Gasoline rationing may be necessary. These events will have a multiplying effect and could cause a real dislocation of our society.

25.3 A DILEMMA?

What must we do now to avoid facing a real crisis in the next decade or two?

First of all, we the citizens as well as our leaders, particularly those in government, must recognize that the days of almost unlimited, inexpensive energy are gone and gone forever. We must use our present supplies wisely and, at the same time, devote much of our resources to developing alternative energy sources. We must recognize that the development of these resources requires great sums of money. Hence, we, the general public, must be prepared, directly or indirectly, to pay heavily for the development of these sources.

Our dilemma is this: How much are we willing to decrease our energy consumption and what are the alternatives which must be considered without a reduction? Many of our energy problems would be minimized if we reduced our total energy demands to those of only 30 years ago, which were approximately 40 percent of our present demands. If we decide to do this, a sizeable portion of our population would have to be shifted back to the farm. Manual labor would be used much more extensively both within our factories and on the outside. Our freedom of movement by car would be greatly restricted. The use of many devices in our homes, such as air conditioners, would have to be restricted. If we decide to make such a reduction in our use of energy, how is it to be done? Will voluntary action be sufficient or will a government edict be necessary?

Assuming that we are not willing to make a drastic reduction in our energy demands, then we must start now to undertake developments which will provide us with significant amounts of energy within the next decade. If this be our decision, several major factors must be considered:

1. Which energy source or sources are the most promising and to which should we devote large resources? Should we expand on our use of coal and nuclear fission? Should we develop solar energy, wind power, geothermal energy, shale oil, ocean thermal gradients or others? How can we decide this question without factual information on full-size units?
2. How can we ensure that we will have sufficient financial resources to make the desired developments? Shall we depend on the federal government to provide the funds and to make the developments? Or should we permit the energy companies to make sufficient profits to undertake these developments? If this be our decision, how can we be certain that the profits from the energy companies are spent wisely for our benefits and not be given in excessive amounts to the stock holders? Above all, are we willing to pay, directly or indirectly, the price that is necessary for the development of additional energy supplies?

3. As we draw sizeable amounts of energy from nature and use this energy we will affect the environment. How much of a compromise must we make between obtaining energy and affecting the environment? How much are we willing to pay to minimize adverse environmental effects?

25.4 OR JUST ANOTHER PROBLEM?

Our crisis cannot be avoided and our dilemma cannot be resolved satisfactorily until we have a detailed, well thought-out, long-range energy program. We may muddle along in some sort of a fashion as Congress, the Executive branch of government, and our industrial leaders make day-to-day decisions. Some of these decisions are helpful but some are contradictory not only of decisions made by other bodies, but, in some cases, made earlier by these groups themselves.

The energy problem is extremely complex. There are so many dilemmas involved that it is essential that the best informed and the most capable individuals work together to formulate a long-range comprehensive continuing energy program for this nation. Then, and only then, can the energy situation be reduced to just another problem. Even then, any such program will not and cannot work without the support of a well-informed citizenry and of understanding leaders in both government and industry.

25.5 CONCLUSION

Energy—a crisis, a dilemma or just another problem? We, as citizens, can answer this question with our interest, our knowledge, our participation, and our guidance for those in positions to make decisions. We must support wise decisions and denounce unwise ones.

SUPPLEMENT

DISCUSSION QUESTIONS

25-1. Do you feel that we will experience a real energy crisis in this decade? Why?

25-2. Discuss the probability of our present energy crisis (if you feel that one does exist) changing to a dilemma in the next two or three decades.

25-3. Discuss the probability of reducing the energy situation to just another problem during the rest of this century.

25-4. How can the energy situation be reduced to just another problem? What role should and will the government, business, and the general public play in improving the energy situation?

25-5. How can the general public be made to realize that the days of abundant and cheap energy are gone forever?

25-6. Table 2-1 indicates that we will demand 115 Quads of energy by 1995 (compared with 78 Quads of energy used in 1979). Do you feel that it is possible to meet this demand without greatly increasing our use of coal and nuclear fission? If so, how?

25-7. If we do not greatly increase our use of coal and nuclear fission do you feel that there will be a crisis in this country by 1995? Why?

References, First Edition

Note: The references listed here were selected from an exceedingly large number of valuable and informative publications in the energy field. Those general publications which are listed give excellent background material for the various aspects of energy discussed in this text. Most of these publications contain extensive additional reference material.

Those publications with the designation * are published by the Government Printing Office and should be available through the Superintendent of Documents. Those publications with the designation † should be available through the Technical Information Center, Oak Ridge, Tennessee.

GENERAL

1. *A National Plan for Energy Research, Development and Demonstrations.* Energy Research and Development Administration. Washington, D.C. 1975.*
2. *A Time to Choose: America's Energy Future.* The Ford Foundation. Ballinger Publishing Co. Cambridge, Mass. 1974.
3. "Energy." *Science.* April 19, 1974.
4. *Energy Alternatives: A Comparative Analysis.* Science and Public Policy Program. University of Oklahoma. Norman, Oklahoma. 1974.*
5. *Energy Basis for Man and Nature.* H. T. Odum and E. C. Odum. McGraw-Hill Book Company. New York, N.Y. 1976.
6. *Energy and Environment.* Edward H. Thorndike. Addison-Wesley Publishing Company. Cambridge, Mass. 1976.
7. *Energy and Environment,* Gary O. Robinette. Kendall/Hunt Publishing Company. 1973.
8. "Energy and the Future." *Science.* 1973.
9. "Energy—Environment Source Book." National Science Teachers Association. Washington, D.C. 1975.
10. *Energy. From Nature to Man.* William C. Reynolds. McGraw-Hill Book Company. New York, N.Y. 1974.

11. *Energy Perspectives*. United States Department of Interior. Washington, D.C. 1975.*
12. "Energy: Use, Conservation and Supply." *Science*. 1974.
13. *National Energy Outlook*. Federal Energy Administration. Washington, D.C. 1976.*
14. *New Energy Technology: Some Facts and Assessments*. H. C. Hottel and J. B. Howard. MIT Press. Cambridge, Mass. 1971.
15. *Project Independence. A Summary*. Federal Energy Administration. Washington, D.C. 1974.*
16. *The 1970 National Power Survey, Part I*. Federal Power Commission. Washington, D.C. 1971.*
17. *The Energy Crisis*. Lawrence Rocks and Richard Runyon. Crown Publishers. New York, N.Y. 1972.
18. *The Nation's Energy Future*. United States Atomic Energy Commission. Washington, D.C. 1973.*
19. *The U.S. Energy Problem*. Volume I, Part A and Part B. Inter Technology Corporation. Warrenton, Virginia. 1971.
20. *Towards Project Interdependence: Energy in the Coming Decade*. Joint Committee on Atomic Energy, United States Congress. Washington, D.C. 1975.*
21. *Understanding the National Energy Dilemma*. Joint Committee on Atomic Energy, United States Congress. Published by The Center for Strategic and International Studies. Washington, D.C. 1973.
22. *United States Energy Through the Year 2000*. Bureau of Mines, U.S. Department of the Interior. Washington, D.C. 1975.

SPECIFIC

BIOMASS AND WASTES

1. *Energy in Solid Waste—A Citizen's Guide to Saving*. Citizen's Advisory Committee on Environmental Quality. Washington, D.C. 1974.*
2. *Energy Recovery from Waste*. U.S. Environmental Protection Agency. Washington, D.C. 1973.
3. *Proceedings of the 1976 National Solid Waste Processing Conference—From Waste to Resource Through Processing*. American Society of Mechanical Engineers. New York, N.Y. 1976.

CONSERVATION OF ENERGY

1. *Citizen Action Guide to Energy Conservation*. Citizen's Advisory Committee on Environmental Quality. Washington, D.C. 1973.*
2. *Energy Conservation Handbook for Light Industries and Commercial Buildings*. U.S. Department of Commerce. Washington, D.C. 1974.

3. *Energy Conservation in New Building Design*. American Society of Heating, Refrigeration and Air-Conditioning Engineers. New York, N.Y. 1975.
4. *In the Bank or Up the Chimney*. United States Department of Housing and Urban Development. Washington, D.C. 1975.*
5. *Retrofitting Existing Housing for Energy Conservation: An Economic Analysis*. National Bureau of Standards, U.S. Department of Commerce. Washington, D.C. 1974.
6. "Save Energy: Save Money." Office of Economic Opportunity. Washington, D.C. 1975.
7. *The Potential for Energy Conservation*. Executive Office of the President. Washington, D.C. 1973.*

FOSSIL FUELS

1. *Coal Resources of the United States*. Geological Survey, U.S. Department of the Interior. Washington, D.C. January 1, 1974.
2. *Final Report, Volumes I and II. Oil and Gas Resources, Reserves and Productive Capacities*. Federal Energy Administration. Washington, D.C. 1975.
3. *Project Independence, Coal*. Federal Energy Administration. Washington, D.C. 1974.*
4. *Survey of Coal Availabilities by Sulfur Content*. The Mitre Corporation for the Environmental Protection Agency. Washington, D.C. 1972.

THE HYDROGEN ECONOMY

1. *The Hydrogen Economy. Proceedings of the Miami Energy Conference*. University of Miami. Coral Gables, Florida, 1974.

GEOTHERMAL ENERGY

1. *Geothermal Energy Resources and Research*. Hearings Before the Committee on Interior and Insular Affairs, United States Senate. Washington, D.C. 1972.*
2. *The Potential for Energy Production From Geothermal Resources*. Committee on Interior and Insular Affairs. Washington, D.C. 1973.*

NEW METHODS OF POWER PRODUCTION

1. *A National Fuel Cell Program*. Power System Division, United Technology. Hartford, Conn. 1975.
2. *Fuel Cells: A Survey*. B. J. Crowe. Computer Science Corporation for National Aeronautics and Space Administration. Washington, D.C. 1973.*
3. *Open Cycle Coal Burning MHD Power Generation*. Office of Coal Research, U.S. Department of Interior. Washington, D.C. 1971.

NUCLEAR ENERGY

1. *Controlled Nuclear Fusion*. Samuel Glasstone. Division of Technical Information, Atomic Energy Commission. Washington, D.C. 1974.†
2. *Nuclear Power and the Environment*. American Nuclear Society. Hinsdale, Ill. 1973.
3. *Nuclear Power Plants*. R. L. Lyerly and Walter Mitchell, III. Division of Technical Information, Atomic Energy Commission. Washington, D.C. 1973.†
4. *Radioactive Wastes*. Charles H. Fox. Division of Technical Information, Atomic Energy Commission. Washington, D.C. 1969.†
5. *Sources of Nuclear Fuel*. Arthur L. Singleton, Jr. Division of Technical Information, Atomic Energy Commission. Washington, D.C. 1968.†
6. *Uranium Enrichment. A Vital New Industry*. Division of Technical Information, Energy Research and Development Administration. Washington, D.C. 1975.†
7. *Worlds Within Worlds. The Story of Nuclear Energy*. Volumes 1, 2, and 3. Isaac Asimov. Division of Technical Information, Atomic Energy Commission. Washington, D.C. 1972.†

OCEANS

1. *Energy From the Oceans: Fact or Fantasy—Conference Proceedings*. The Center for Marine and Coastal Studies—North Carolina State University. Raleigh, N.C. 1976.

SOLAR ENERGY

1. *Applied Solar Energy*. A. B. Meinel and M. P. Meinel. Addison-Wesley Publishing Company. Cambridge, Mass. 1976.
2. *Direct Use of the Sun's Energy*. Farrington Daniels. Yale University Press. New Haven, Conn. 1964.
3. *Solar Energy for Man*. B. J. Brinkworth. John Wiley & Sons. New York, N.Y. 1972.
4. *Solar Heating and Cooling*. J. F. Kreider and Frank Kreith. McGraw-Hill Book Company. New York, N.Y. 1975.
5. "Solar Heating Equipment." *Popular Science*. March, 1975.
6. *Solar Heating for Heating and Cooling of Buildings*. A. R. Patton. Noyes Data Corporation. Park Ridge, N.J. 1975.
7. *Solar Energy Thermal Processes*. J. A. Duffie and W. A. Beckman. John Wiley & Sons. New York, N.Y. 1974.
8. "Solar Water Heaters You Can Build or Buy." *Popular Science*. May, 1976.

TAR SANDS

1. *Syncrude. Mining the Athabaska Tar Sands*. The Orange Disc, Gulf Oil Companies. Pittsburgh, Penn. 1975.

WIND POWER

1. *Proceedings of the Second Workshop on Wind Energy Conversion Systems*. Sponsored by the National Science Foundation. Washington, D.C. 1975.*

There are many technical periodicals which contain excellent articles with various aspects of energy. Some of these publications are

1. *Combustion*
2. *Electrical World*
3. *Mechanical Engineering*
4. *Nuclear Science and Engineering*
5. *Popular Science*
6. *Power*
7. *Science*
8. *Scientific American*

Many governmental agencies develop and publish information dealing with various aspects of energy. Some of these agencies are

1. Energy Research and Development Administration
2. Environmental Protection Agency
3. Federal Energy Administration
4. U.S. Bureau of Mines
5. U.S. Department of Commerce
6. U.S. Department of Housing and Urban Development
7. U.S. Department of Interior
8. U.S. Federal Power Commission
9. U.S. Geological Survey.

Selected References, Second Edition

GENERAL

1. "Data." *The 1979 Annual Report to Congress, Volume Two*. U.S. Department of Energy, Energy Information Administration. Washington, D.C.*
2. "Energy—A Special Report in the Public Interest. Facing up to the Problem, Getting Down to Solutions." *National Geographic*. February, 1981.
3. *Energy Information Digest—Basic Data on Energy Resources, Reserves, Production, Consumption and Prices*. Congressional Research Service, Library of Congress. Washington, D.C. July, 1977.*
4. *U.S.A.'s Energy Outlook 1980-2000*. Exxon Company, Public Affairs Department. Houston, Texas. December, 1979.
5. "Synopsis of Energy Facts and Projections." *The 1978 Annual Report to Congress*. U.S. Department of Energy, Energy Information Administration. Washington, D.C.*
6. *Research Information Packages*. Smithsonian Science Information Exchange, Suite 300, 1730 M Street, N.W., Washington, D.C. 20036. Approximately 400 packages available for purchase in all areas of energy research. Each package includes the names of principal researchers and a technical summary.

SPECIFIC

BIOMASS AND WASTE

1. *Energy From Biomass and Solid Wastes: Prospects and Constraints*. Science Policy Research Division, Congressional Research Service, Library of Congress. Washington, D.C. June, 1980.*
2. *Fuels from Biomass, Technology and Feasibility*. Edited by J. S. Robinson. Noyes Data Corporation. Park Ridge, N.J. 1980.

* Government publications may be obtained from the Superintendent of Documents, U.S. Government Printing Office, Washington, D.C. 20402.

3. *Biomass Production and Utilization*. Elizabeth Price and Paul Cheremisonoff. Ann Arbor Publishers. Woburn, Massachusetts. 1980.
4. *Large and Small Scale Ethyl Alcohol Manufacturing Processes From Agricultural Raw Materials*. Edited by J. K. Paul. Noyes Data Corporation. Park Ridge, N.J. 1980.
5. "Alcohol Fuels Project." *Mechanical Engineering*. Volume 102, No. 10. October, 1980.
6. *Methanol Technology and Applications in Motor Fuels*. Edited by J. K. Paul. Noyes Data Corporation. Park Ridge, N.J. 1978.
7. *Biomass Applications Technology and Production*. N.P. Cheremisonoff, P.N. Cheremisonoff, F. Ellenbush, Marcel Dekker, Inc. 1980.

ENVIRONMENT

1. "Acid Lakes From Natural and Anthropogenic Causes." *Science*. Vol. 211, No. 4481. January 30, 1981.
2. "Acid Rain." *Combustion*. Vol. 52, No. 4. October, 1980. p. 12.
3. "Acid Rain Controversy." *Combustion*. Vol. 52, No. 4. October, 1980. p. 17.
4. "Detecting Climatic Changes Due to Increasing Carbon Dioxide." R. A. Madden and V. Ramanathan. *Science*. Vol. 209, No. 4458. August 15, 1980. p. 783.

FOSSIL FUELS

1. *Coal Information Sources and Data Bases*. Carolyn C. Bloch. Noyes Data Corporation. Park Ridge, N.J. 1980.
2. *Coal Resources, Characteristics and Ownership in the U.S.A.*. Edited by R. Noyes. Noyes Data Corporation. Park Ridge, N.J. 1978.
3. "Enhanced Recovery of Crude Oil." Todd M. Doscher. *American Scientist*. Vol. 69, No. 2. April, 1981. p. 193.
4. *Unconventional Natural Gas Resources, Potential and Technology*. Edited by M. Satriana. Noyes Data Corporation. Park Ridge, N.J. 1980.
5. "U.S. Coal Gasification Program, Progress and Projects." C. L. Miller. *Mechanical Engineering*. Vol. 102, No. 8. August, 1980. p. 34.

GEOTHERMAL ENERGY

1. "A Status Report—Hot Dry Rock Geothermal Energy." G. J. Nunz. *Mechanical Engineering*. November, 1980. p. 26.
2. "Eastern Geothermal Resources: Should We Pursue Them?" J. E. Tillman. *Science*. Vol. 210, No. 4470. November 7, 1980. p. 595.
3. "Geothermal Energy: An Assessment." R. W. Potter, II. *Mechanical Engineering*. Vol. 103, No. 5. May, 1981. p. 20.
4. *Geothermal Energy as a Source of Electricity. A Worldwide Survey of the Design and Operation of Geothermal Power Plants*. Ronald Di Pippo, Assistant Secretary for Resource Applications, U.S. Department of Energy, Division of Geothermal Energy. Washington, D.C. January, 1980.*

5. "Geothermal Energy—Recent Developments." Edited by M. J. Collie. Noyes Data Corporation. Park Ridge, N.J. 1978.

HYDROPOWER

1. "Economic Development of Small Hydro." C. A. Fritz Neuman. General Electric Company. Schenectady, N.Y. 1979. Reprint from *Electric Forum*. No. GER 3187.
2. "Small and Micro Hydroelectric Power Plants." Edited by Robert Noyes. Noyes Data Corporation. Park Ridge, N.J. 1980.
3. "Will Your Small-Hydro Development Be a Liquid Asset or Only a Liability?" William O'Keefe. *Power*. Vol. 125, No. 1. January, 1981. p. 75.

NEWER METHODS OF POWER PRODUCTION

1. "A Comparison of Power Plants for Cogeneration of Heat and Electricity." R. Kehlhofer. *Combustion*. Vol. 52, No. 8. March, 1981. p. 22.
2. "Molten Carbonate Fuel Cells." *Mechanical Engineering*. Vol. 102, No. 8. August, 1980. p. 22.
3. "Purpa 210: New Life for Cogenerators." Douglas J. Turtle. Edited by Paul S. Baur. *Power*. Vol. 125, No. 1. January, 1981. p. 78.
4. "Solar-Powered Liquid-Metal MHD Power Systems." E. S. Pierson, H. Branover, G. Fabris, C. B. Reed. *Mechanical Engineering*. Vol. 102, No. 10. October, 1980. p. 32.
5. "The Combined Cycle Power System Using Synthetic Fuels." F. Robson. *Mechanical Engineering*. Vol. 102, No. 12. November, 1980. p. 35.
6. "MHD Power Generation: Not Now, But Ever?" *Compressed Air*. October, 1981. p. 10.

NUCLEAR ENERGY

1. "Do We Need the Breeder?" Edward Edelson. *Popular Science*. Vol. 217, No. 5. November, 1980. p. 72.
2. "Clinch River Progress Resport." *Mechanical Engineering*. Vol. 102, No. 8. August, 1980. p. 22.
3. *Nuclear Power Quick Reference II*. General Electric Company, Nuclear Energy Group. San Jose, California. 1979.
4. "Putting Nuclear Energy Into Perspective." Sam A. Wenk. *Mechanical Engineering*. Vol. 102, No. 13. December, 1980. p. 42.
5. "Slow Breeder Makes Its Own Fuel." James G. Busse. *Power*. Vol. 212, No. 4. April, 1978. p. 89.
6. "The Future of Nuclear Power in the U.S." J. C. Deddens. *Mechanical Engineering*. Vol. 103, No. 2. February, 1981. p. 27.
7. "The Hunt for Giant Uranium Deposits." Eric S. Cheney. *American Scientist*. Vol. 69, No. 1. January-February, 1981. p. 37.
8. "The Legacy of TMI-2 for the Nuclear Power Plant Operations System." W. L. Livingston. *Combustion*. Vol. 52, No. 5. November, 1980. p. 10.

9. "The Next Step in Fusion: What It Is and How It Is Being Taken." John F. Clark. *Science*. Vol. 210, No. 4473. November 28, 1980. p. 967.

OCEANS

1. "S.S. Converter." *Mechanical Engineering*. Vol. 102, No. 12. November, 1980. p. 20.
2. *Ocean Thermal Energy Conversion (OTEC) Program*. Division of Solar Technology, U.S. Department of Energy. Washington, D.C. 1978. (Available from National Technical Information Service, U.S. Department of Commerce, 5285 Port Royal Road, Springfield, Virginia 22161.)
3. "Tapping the Oceans' Vast Energy with Undersea Turbines." P. B. S. Lissaman. *Popular Sciences*. Vol. 217, No. 3. September, 1980. p. 72.
4. "Tapping Sun-Warmed Ocean Water for Power." Beverly Karplus Hartline. *Science*. Vol. 209, No. 4458. August 15, 1980. p. 794.

OIL SHALE AND TAR SANDS

1. *An Assessment of Oil Shale Technology*. Office of Technology Assessment, United States Congress. Washington, D.C. June, 1980.*
2. *Oil Shale and Tar Sands Technology: Recent Developments*. M. W. Ranney. Noyes Data Corporation. Park Ridge, N.J. 1979.
3. "Shale Oil Sponsors." *Mechanical Engineering*. Vol. 102, No. 13. December, 1980. p. 20.
4. *U.S. Tar Sands Oil Forecasts (1985-1995)*. Assistant Administrator for Applied Analysis, Energy Information Administration, U.S. Department of Energy. Washington, D.C.*

SOLAR ENERGY

1. "And Now It Is Sun Power." *Time*. Vol. 115, No. 8. February 25, 1980. p. 39.
2. "Solar Ponds." *Mechanical Engineering*. Vol. 102, No. 10. October, 1980. p. 20.
3. *Passive Solar Energy Design and Materials*. Edited by J. K. Paul. Noyes Data Corporation. Park Ridge, N.J. 1979.
4. *Design of a Solar Space Heating System*. W. F. Bessler and H. Jaster. Power Systems Laboratory, General Electric Company. Schenectady, N.Y. June, 1977.
5. "Principles and Applications of Solar Energy." Paul N. Cheremisinoff and Thomas Regino. Ann Arbor Science. Woburn, Mass. 1978-79.
6. *Principles of Solar Engineering*. Kreith and Kreider. McGraw Hill Publishing Company. New York, N.Y. 1980.
7. *Solar Cells for Photovoltaic Generation of Electricity – Materials, Devices and Applications*. Marshall Sittig. Noyes Data Corporation. Park Ridge, N.J. 1979.

8. *Solar Heating and Cooling: Recent Advances*. J. K. Paul. Noyes Data Corporation. Park Ridge, N.J. 1977.
9. *Solar Heating and Cooling of Buildings*. Robert P. Quellette and Paul N. Cheremisenoff. Ann Arbor Science. Woburn, Mass. 1981.
10. "Solar Photovoltaic Power for Residential Use." B. Hammond. *Mechanical Engineering*. Vol. 102, No. 3. December, 1980.
11. "Taking the Mystery Out of Passive Solar Design." Benjamin T. Rogers and Daniel Ruby. *Popular Science*. Vol. 218, No. 2. February, 1981. p. 76.
12. *Photovoltaics Solar Electric Power Systems*. Solar Energy Research Institute. Golden, Colorado. February, 1981. Prepared by PRC Energy Analysis Company, McLean, Virginia.*
13. *Solar Collector Manufacturing Activity, 1978*. Energy Information Administration, U.S. Department of Energy. Washington, D.C.*

WIND POWER

1. "Blue Ridge Electric Takes a Close Look at Wind Power." Edited by Sheldon Strauss. *Power*. Vol. 125, No. 1. January, 1981. p. 56.
2. "Giant-Windmill Array." Ben Kouvar. *Popular Science*. Vol. 218, No. 2. February, 1981. p. 89.
3. *Wind Power: Recent Developments*. Edited by D. J. DeRenzo. Noyes Data Corporation. Park Ridge, N.J. 1979.
4. *Wind Energy Conversion*. Environmental and Resource Assessments Branch, Division of Solar Energy, Research and Development Administration. Washington, D.C. March, 1977.*
5. *Wind Power and Windmills*. Science and Educational Administration, U.S. Department of Agriculture. Washington, D.C. March, 1980.*
6. "World's Biggest Wind Machine Is a One-Armed Monster." *Popular Science*. January, 1981. p. 83.

APPENDICES

Appendix A

Energy Concepts and Energy Use

A.1 ENERGY, FORCE, PRESSURE, TEMPERATURE

Although most of us may have a general concept of what *energy* is, it is essential that we fully understand its nature in order to comprehend the use of various sources in meeting our energy needs.

What is energy? A common definition of energy is the capacity for doing work. This definition is not very satisfactory, particularly if we consider work to be mechanical. Energy, per se, cannot be seen; it cannot be felt. It can be conceived of only by the results which it produces. Thus, we may say that energy produces physical and/or chemical changes in material objects as a whole or in their component parts. Energy is required to change the physical location of an object, such as lifting a book from the floor to a table. Energy is required to deform a structure. Energy is required to heat a pan of water. Energy is delivered up by a storage battery as a result of chemical action. Energy must be supplied to restore the chemicals in the battery to their original state. Energy is required to move a car. Energy is required to vaporize water in the oceans. Part of this energy is released when the vapor condenses to form rain or snow. More energy is made available when the rain forms a river and is used to produce hydroelectric power.

Before proceeding with a more complete concept of energy, it is desirable to establish concepts of several quantities which are involved, directly or indirectly with energy. We will consider such quantities as heat, work, force, and temperature.

Although we all have a general concept of what is meant by the term *force*, a precise definition is desirable. Sir Isaac Newton provided us means of making such a definition with his force-mass-acceleration equation.

$$F = ma \qquad \text{(A-1)}$$

Equation A-1 does not define force as such but it relates force F, to mass m, and acceleration a. The mass of an object is the amount of matter that composes the object. This amount can be obtained, numerically, by referring, indirectly, to a standard mass. The standard mass is a precise amount of a platinum-iridium alloy located in Sevres, France. This standard mass has been designated as the kilogram mass (or 1,000 grams mass). The mass of any other object may be obtained by comparing it to a known standard, using a balance (see Figure A-1). If the balance arm is level, the mass determined must be equal to that of a standard. The pound mass, denoted by lb_m, is defined as 0.45359 times the kilogram mass.

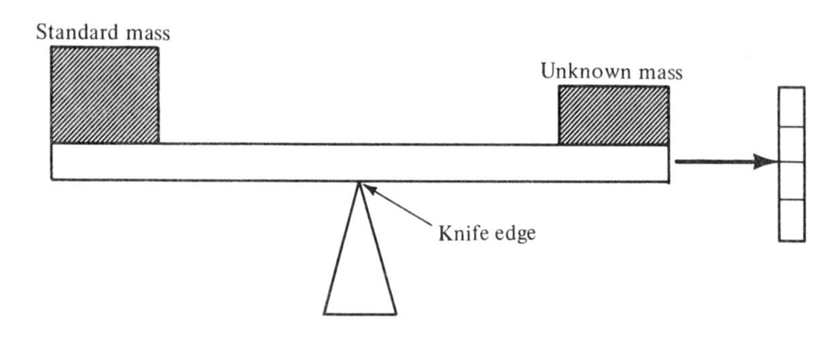

Figure A-1. Mass determination

If, in Equation A-1, m is expressed in kilogram and a is meters per second per second, the unit of force is established as the Newton. Thus, the Newton is that force which will accelerate a kilogram mass at the rate of 1 meter per second per second (m/s^2). In the engineering or English system of units,* the common unit of force is the pound force (lb_f) and, as noted above, the common unit of mass is the pound mass (lb_m). This is most unfortunate, since the pound force will accelerate the pound mass at a rate of 32.1740 feet per sec per sec (ft/sec^2). To make equation A-1 numerically and dimensionally sound, a dimensional constant, g_c, is introduced. The value of g_c is 32.1740 lb_m ft/lb_f sec^2. Then,

$$F' = \frac{ma}{g_c} \tag{A-1a}$$

When one pound force and one pound mass are involved Equation A-1a becomes

$$1 \text{ lb}_f = \frac{1 \text{ lb}_m}{32.174 \text{ lb}_m \text{ft/lb}_f \text{sec}^2} \times 32.174 \text{ ft/sec}^2$$

---

* As stated in the Preface, the engineering or English system of units is used throughout this text since most people are more familiar with it than with the SI system. However, SI system is discussed in detail in Appendix B and methods of conversion between the two systems are given.

Weight is defined as the force of gravity acting on a body. Since, at standard conditions,* the force of gravity acting on one pound mass is one pound force, the weight of a pound mass at these conditions is one pound force. Equation A-1a may be written as

$$W=\frac{m}{g_c}g \qquad \text{(A-1b)}$$

where W *is the weight in* lb_f
g is the gravity acceleration at the given location, ft per sec per sec (ft/sec²)

At higher elevations, the value of g is somewhat lower than its sea level value of 32.174 ft/sec². However, for most engineering problems, the elevation involved is not sufficiently high to make a significant reduction in g. Assuming that g and g_c are numerically equal, it may be seen that the weight of an object, expressed in lb_f and the mass of the object in lb_m, are numerically equal and may be used as such in most engineering problems.

Pressure is defined as a force per unit area. Normally, pressure is used in connection with a fluid (gas, vapor or liquid). The common units of pressure are pounds per square inch (psi). Pressures commonly are measured by use of a pressure gauge, which measures pressures above or below that of the atmosphere. These pressures are expressed in pounds per square inch gauge (psig). Thermodynamically speaking, the pressure which must be used is the absolute pressure, which is the pressure in excess of that of a perfect vacuum. It may be designated as pounds per square inch absolute (psia). Using the observed gauge pressure,

$$P_{absolute}=P_{gauge}+P_{atmosphere} \qquad \text{(A-2)}$$

$P_{atmosphere}$ frequently is known as the barometric pressure.

We use our sense of touch to tell us if an object is hot, warm, cool or cold. Such a concept is not useable, however, in energy calculations. It is difficult to give a precise definition of *temperature*. Rather than become too involved, we will introduce an approximate approach which should be satisfactory for our purpose. The properties of various substances vary with temperature. A readily measurable property is volume. A thermometer consists of a small bore (capillary) tube with a reservoir at its base. The thermometer is partially filled with a liquid, such as mercury. Two reference points are established. One of these is the freezing point of pure water (the ice point)† and the other is the boiling point of pure water (the steam point), both at standard atmospheric pressure of 14.696 psia. Two scales of temperature are in use. One of these, the Fahrenheit, designates the temperature at the ice point as 32°F and the steam

* Sea level, 45 degrees North latitude.

† The ice point has been replaced as standard by the triple point, at which water, ice, and water vapor exist in equilibrium conditions. It is very close to the ice point.

point as 212°F. The other scale, the Centigrade or Celsius scale, designates the ice point as 0°C and the steam point as 100°C. Temperature may be converted from one scale to another as follows:

$$°F = °C \times \frac{212-32}{100-0} + 32 = °C\frac{180}{100} + 32 \tag{A-3a}$$

$$°C = (°F - 32)\frac{100}{180} \tag{A-3b}$$

The Fahrenheit and Celsius scales may be extended to measure temperatures below the ice point and above the steam point. Additional reference temperatures, such as the melting temperature of pure sulfur, have been established.

Although we now have means of measuring temperature, we have not presented a concept of it. Some concept of temperature may be established by considering an ideal gas.* By virtue of their velocities, the molecules of the gas possess kinetic energy. Temperature is said to be a measure of this kinetic energy. If a thermometer is placed in a high temperature gas, the gas molecules, by virtue of their high velocities, will transfer energy to the glass of the thermometer. In turn, the glass will transfer energy to the liquid in the thermometer, causing it to expand. Thus the reading on the thermometer is a function of the kinetic energy of the gas molecules and its temperature.

As we cool the gas, the kinetic energy of the molecules and hence the temperature of the gas decreases. If we imagine that the gas does not liquify as we cool it but always remains as an ideal gas, ultimately as it is cooled, the molecules of the gas will cease to move and hence would possess no kinetic energy. The temperature at this point is said to be the absolute zero of temperature. Of course, all known gases liquefy before this absolute level of temperature is reached. To account for this, consider a box (having constant volume) containing an ideal gas. By virtue of their kinetic energies, the gas molecules exert a pressure. As the gas is cooled the kinetic energy and hence the gas pressure decreases. The pressure is plotted against temperature in Figure A-3. This plot is extrapolated to zero pressure. At zero pressure, the molecular velocity and hence the temperature becomes zero. This is the absolute zero of temperature.

Temperatures measured above absolute zero, using Fahrenheit degrees, are expressed in degrees Rankine, °R. Using Celsius degrees, absolute temperatures are expressed in degrees Kelvin, °K. Conversions to absolute temperatures may be made as follows:

$$°R = °F + 459.67 \tag{A-4a}$$

$$°K = °C + 273.15 \tag{A-4b}$$

---

* Consider the molecules of a gas moving with random velocities within a container. If the size of the molecules and forces existing between the molecules can be neglected, we say that the gas is ideal.

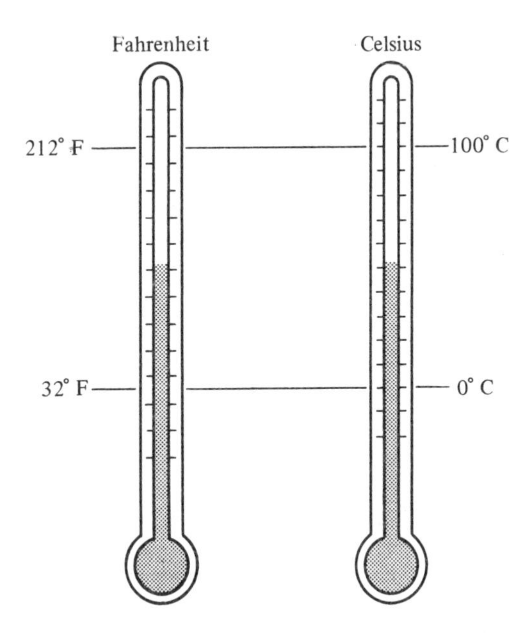

Figure A-2. Fahrenheit and Celsius thermometers

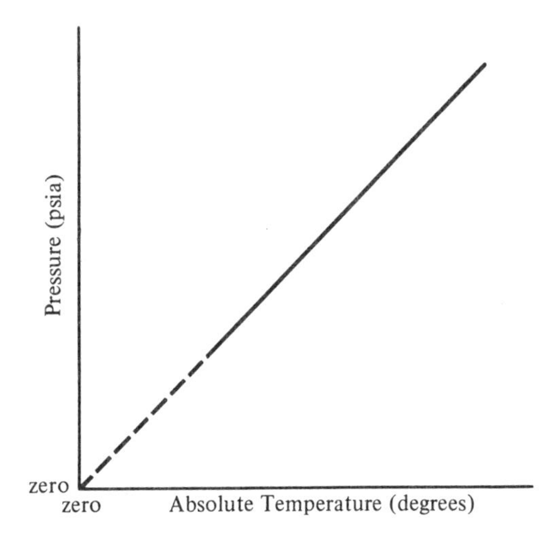

Figure A-3. Absolute temperature

A.2 FORMS OF ENERGY

Energy has many forms, such as heat, work, electrical energy, chemical energy, and internal energy. In some cases we may deal with energy which is being transferred from one object to another. In other cases we may be concerned with energy which is possessed by the objects. If we are to avoid confusion, it is essential that we be very careful to specify which type of energy we are talking about. It is also essential that we specify the objects which we are considering. For example, if we are talking about what is taking place within the cylinders of an automotive engine while our reader is thinking about the entire engine, there will not be a meeting of minds. To avoid this situation, it is common practice to specify a *system*. This is done by imaging an envelope to enclose all objects under consideration. Once we specify a system, we may proceed to the consideration of the energies involved without causing confusion on the part of others.

Energy may enter or leave a system. It may also be possessed by a system or by any physical object. Care must be taken to distinguish between energy possessed by a system and the energy transferred to or from the system. Care must also be taken not to use various forms of energy interchangeably. For example, it is common practice to rate a nuclear reactor, say, as 1,000,000 kW. This is very misleading since the electrical output of a power plant using this reactor may be only 320,000 kW. This confusion may be avoided, if care is taken to specify the nature of the energy. Thus we should say that the output of the reactor is 1,000,000 kW_t (kW thermal) and the output of the plant is 320,000 kW_e (kW electrical).

Three types of energy which are transferred are (1) heat, (2) work, and (3) electrical energy. Each of these is associated with a driving force. The work referred to here is mechanical work, which will be discussed shortly. Some authors lump mechanical work and electrical work together under the general term of work. For the purposes here, however, it is desirable to treat them separately.

For *heat* the driving force is temperature. When a *temperature difference* exists between two points, heat will be transferred between the two points, provided that a heat transfer path exists. Heat may be expressed in either Btu (British thermal units) or in calories. For the purposes of this book, a Btu is defined as the amount of heat required to increase the temperature of 1 lb_m of water by 1°F at "room" conditions. The calorie, then, is the amount of heat required to increase the temperature of 1 gram of water 1°C. The relationship between the Btu and calorie is:

$$1 \text{ Btu} = \frac{453.59}{1.8} = 252 \text{ calories} \tag{A-5}$$

It is very difficult to measure heat as it is being transferred. Commonly, heat is determined by the results which it produces. Normally, when heat is added to a substance, it produces a temperature rise of that substance. The term *specific heat c* refers to the amount of heat which must be added to increase the temperature of a unit mass of the material by one degree. Thus

$$Q = mc(T_2 - T_1) \tag{A-6}$$

where Q = heat

Equation A-6 assumes that the specific heat is independent of temperature. When the specific heat varies significantly with temperature it is necessary to relate specific heat and temperature and to put Equation A-6 in the form of a differential equation.

Under most conditions for solids and liquids, changes in pressure do not significantly affect the specific heat and hence it is not necessary to specify the conditions during heat addition. On the other hand, for gases and vapors, changes in pressure during heat added will affect the specific heat significantly. Two common processes for heat addition to gases and vapors are the constant pressure process and the constant volume process. For the constant pressure process the specific heat is designated as c_p. For the constant volume process it is c_v.

For *work*, the driving force is a mechanical force. When a force causes the movement of an object, work is done (see Figure A-4). This work equals the product of the force, measured in direction of motion, and the distance moved. Thus

$$W = F \times d \tag{A-7}$$

where W = work
F = force in direction of motion
d = distance moved

The relation between heat and work is given as

$$1 \text{ Btu} = 778.16 \text{ ft lb}_f \tag{A-8}$$

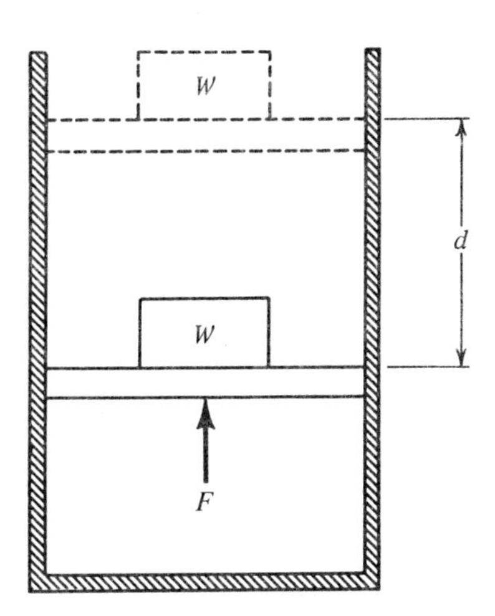

Figure A-4. Work

For *electrical energy*, the driving force is electrical potential or voltage. Where a *voltage differential* exists between two points, *electrical energy* will be transferred between the two paths, provided that a path exists for a flow of electrical energy. It is common practice to express electrical power in terms of the watt. The watt equals the product of the voltage and amperage. The watt is a relatively small amount of power. For common every day use, the kilowatt (kW) is preferred. The kilowatt equals 1,000 watts. For very large amounts of electrical power the the megawatt (MW) is used. A megawatt equals 1,000,000 watts or 1,000 kilowatts.

The energy contained within an object at rest is known as *internal energy, U*. This energy is made up of the energy possessed by the molecules by virtue of their velocities* and their attractive forces. It also includes the energy of the atoms as a whole. When there is a rearrangement of the atoms in the molecules, as in a chemical reaction, the energy released is known as *chemical energy*. Internal energy also includes the energy of the component particles of the atom—the energy of the electrons, the protons, and the neutrons. When there is a change in the structure of the atoms of an object, the energy released is known as *nuclear energy*.

An object in motion possesses, in addition to its internal energy, both kinetic and potential energy. Assuming that we intend to use both the lb_m and the lb_f, the equation for kinetic energy is

$$K.E. = \frac{mV^2}{2g_c} \tag{A-9}$$

where K.E. = kinetic energy in ft lb_f
m = mass in lb_m
V = velocity in ft/sec
$g_c = 32.174\ lb_m ft/lb_f sec^2$

Potential energy may be evaluated from the following equation:

$$P.E. = \frac{g}{g_c} mZ \tag{A-10}$$

where P.E. = *potential energy, ft* lb_f
g = gravity acceleration for given location, ft/sec^2
Z = difference in elevation between that of the object and that of an arbitrarily chosen reference point

Note that the potential energy calculated from Equation A-10 is not an absolute quantity but simply the energy in excess of that at the reference point. Normally, we are not concerned with the absolute amount of energy possessed by a system but rather with the change in its energy as it changes state. We can determine this change in energy by evaluating the energy of the system, both initial and final, in excess of that at some reference point.

* As stated earlier, temperature is associated with the kinetic energy of the molecules. Much care must be taken to avoid thinking of this energy of the molecules as heat. Otherwise there may be much confusion between the molecular kinetic energy and the energy transferred as the result of a temperature difference.

It requires energy to move a fluid across a reference plane, such as in Figure A-5. As the fluid crosses the reference plane, it carries this energy with it. This energy, for a given amount of the fluid, equals the product of the volume of the fluid and its pressure, pV. This energy of a flowing fluid is combined with its internal energy and its termed *enthalpy* H.

$$H = U + pV \tag{A-11}$$

Thus the total energy possessed by a *flowing fluid* is the summation of its enthalpy, its kinetic energy and its potential energy. The total energy *possessed by an isolated system* (i.e., a system which is completely isolated), is the summation of its internal energy, its kinetic energy and its potential energy. Of course, if the system is at rest, its potential energy does not change and it does not possess kinetic energy.

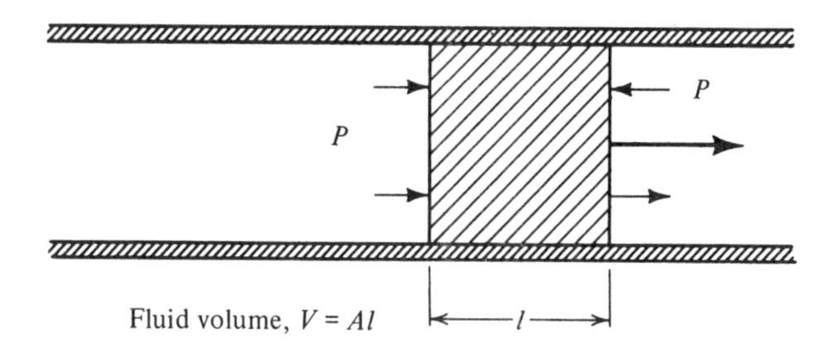

Figure A-5. Energy of flow

Power is defined as the time rate of doing work. it is the work done per unit time. The common unit of power is the horsepower. A horsepower is defined as work done at the rate of 33,000 ft lb per min. Thus the horsepower,

$$\text{hp} = \frac{\text{Work, ft lb}}{(\text{Time in min})\ 33{,}000} \tag{A-12}$$

33,000 ft lb per min is equivalent to 550 ft lb per sec. When work is expressed in Btu equivalent, it is to be noted that 42.4 Btu per min is equivalent to one horsepower.

When dealing with electrical energy, it is common practice to express energy in kilowatt hours (kWh). A kilowatt hour is the power produced at the rate of one kilowatt for one hour. When electrical energy is expressed in terms of Btu,

$$\text{kWh} = \frac{\text{Btu per hr}}{3{,}413} \tag{A-13}$$

where 3,413 = Btu equivalent to one kWh.

The conversion of kilowatt hours to horsepower is

$$1\ \text{hp} = 0.746\ \text{kWh} \tag{A-14}$$

A.3 THE FIRST LAW OF THERMODYNAMICS

Energy may be transferred from one system to a second system, provided that driving forces exist to cause this transfer. A potential may exist to transform energy from one form into a second form. The first law of thermodynamics is simply an accounting of the energies transferred or transformed. It states that, for energy transfer and transformation processes, the total energy in all forms before transfer and/or transformation must equal the total final energy in all forms. The first law is, in reality, a restatement of the law of conservation of energy. It is to be noted that in nuclear processes, both fission and fusion, matter may be transformed into energy or vice versa, with a precise relationship existing between the matter and the energy involved. This fact must be recognized when applying the first law to processes involved in nuclear reactors.

As applied to a system, the first law may be stated as

$$\text{Energy of fluids entering} + Q = \text{change in energy of the system} + \text{energy of fluids leaving} + W_m + W_e \tag{A-15}$$

where Q = the heat added
W_m = mechanical work delivered from the system
W_e = electrical work delivered from the system

See the previous section for the evaluation of the energy of flowing fluids and of the system. It should be noted that all energies must be evaluated for the same time interval. Equation A-15 is a general equation. When fluids do not enter or leave the system, their energies in Equation A-15 become zero.

A.4 THE SECOND LAW OF THERMODYNAMICS

The second law of thermodynamics involves so many aspects of energy transformations that it is very difficult to formulate a simple and yet all inclusive statement of the law. Perhaps the most pertinent aspect of the second law to be discussed here is the limitation of our ability to transform heat into work. This concept is based on the work of Sadi Carnot, published in 1824. In reality this concept may be envisioned by assuming that heat is taken from a high temperature *source* and added to a system. Within the system some, but not all, of the heat may be turned into work. To have a continuing heat-work transformation process, the heat not turned into work must be rejected to a low temperature *sink*.

The term *thermal efficiency* is used to designate how efficient the system is in transforming heat into work. Thus

$$\eta_t = \frac{W}{Q_s} \tag{A-16}$$

where η_t = thermal efficiency
W = work delivered
Q_s = heat supplied

Carnot, in reality, stated that the thermal efficiency is related to the temperatures of the source and the sink. This relationship can be established by use of the Carnot cycle.* It has been shown that the efficiency of transforming heat into work by use of the Carnot cycle is

$$\text{Carnot } \eta_t = \frac{T_{source} - T_{sink}}{T_{source}} \tag{A-17}$$

It should be noted that the temperature to be used in Equation A-17 is the absolute temperature.

All of the processes used in the Carnot cycle are ideal ones (i.e., no losses such as friction). It has been proved that it is impossible for any other cycle to produce an efficiency higher than that of the Carnot. Thus, for given source and sink temperatures, the maximum possible thermal efficiency is that given by Equation A-17. Because of losses, such as friction, the thermal efficiency of actual heat engines are lower than that of the Carnot cycle, being much lower in many cases.

In a sense, Equation A-17 is a negative concept. It states that we cannot even in the ideal case turn all of the heat into work. Negative concepts are most difficult to prove. Two facts should be recognized, however. In the 158 years since Carnot stated his principle, no one has even been able to conceive any method, even an ideal one, in which the thermal efficiency can exceed that given by Equation A-17. In all of our experiences with actual heat-to-work devices, the efficiency of heat-to-work transformation increases, other things being equal, as the temperature difference between the source and sink increases. Thus our experience tells us that we need an infinite temperature difference to transform all of the heat supplied into work.

Although we cannot transform all heat input into work, the converse is not true. We can turn all work input into heat. This is particularly true when we encounter excessive frictional resistances to motion. Similarly, it is possible to transform all electrical energy into heat, such as when an electric current is passed through a resistance. In the ideal case, we can transform all (mechanical) work into electrical energy and also electrical energy into work. As we minimize losses, we can approach complete transformations of work and electrical energy.

In actual processes involving work and electrical energy, some of these energies deteriorate into heat. In general, we produce both work and electrical energy to perform useful tasks for us. The deterioration of these energies into heat defeats the purpose for which these energies were produced. Resistance,

---

* A cycle is composed of a series of processes, at the end of which a substance (normally a gas or vapor) ends up in its original condition of temperature and pressure. For the Carnot cycle, as normally considered, a gas receives heat from the heat source at a constant gas temperature, as it expands and does work. Next the gas expands and produces more work without any heat transfer until its temperature drops to that of the heat sink. The gas is now compressed at constant temperature giving up heat to the heat sink. Finally, the gas is compressed without any heat being transferred until it reaches its original temperature and pressure.

such as friction and electrical resistance, prevent us from transferring work and electrical energy without deterioration and in using them in an efficient manner. We also dissipate our ability to do work when our processes are uncontrolled ones. For example, water gushing from a reservoir cannot deliver the same amount of work as if it were to be passed through a turbine in a controlled manner. A gas under high pressure behind a piston in a cylinder will perform work when the piston is allowed to move in a controlled manner. If, however, the piston is allowed to fly upward in an uncontrolled manner, the ability of the gas to do work is lost.

When heat is transferred from a high temperature source to a system at a much lower temperature, much of the potential of the heat for doing work is lost. Consider, for example, removing 100 Btu of heat from a source whose temperature is 1,000°R and adding this heat to a system whose temperature is 600°R. Assume a heat sink is available whose temperature is 500°F. Using Equations A-16 and A-17, the theoretical work which could be produced originally from the 100 Btu equals

$$100\frac{1{,}000-500}{1{,}000}=50 \text{ Btu}$$

The theoretical work potential of the 100 Btu added to the system at 600°R is

$$100\frac{600-500}{600}=16.67 \text{ Btu}$$

Thus two-thirds of the potential of the heat for doing work has been lost as a result of using this large temperature difference to transfer heat.

In general, we must avoid friction, turbulence, and unbalanced driving forces (mechanical force, temperature, and electromotive force) when transferring energy if we are to avoid losses of our ability to do work. As we minimize these factors causing losses of work or work potential, we approach what is termed the reversible processes.* We may define a reversible process as one without friction and turbulence and one in which the energy driving forces are balanced.

No actual process is reversible. As such, every actual process is irreversible and some of the potential for doing work is lost. Some processes are close to being reversible; others are far from being reversible. Thus, in some processes the loss of potential for doing work is small but in others it is high. A thermodynamic property, known as entropy, is used to measure the loss of the potential for doing work. Entropy will not be defined or used in this book but is mentioned here for the reader to know that in general entropy changes are related to irreversibility.

---

* When the unbalanced driving forces become zero, there will be no losses of the work potential. Hence the process can be reversed and the system can be restored to its original state without outside aid.

A.5 HEAT AND HEAT ENGINES

At present, with the exception of hydroelectric power, substantially all of the energy* which we use is either for heating or for work, either mechanical or electrical. This energy is obtained by burning fossil fuels or from nuclear energy.

For the average residential oil burning furnace, perhaps only 65 percent of the energy of the fuel is delivered to the house. (We say that the efficiency is 65 percent). There are many losses involved in the operation of the furnace. If all of these losses could be eliminated, then all of the energy of the fuel would be delivered to the house, thus achieving an efficiency of 100 percent. The same statements may be made whenever fuels are burned for heating effects.

On the other hand, as stated earlier in the discussion of the second law, it is impossible, even if all losses were to be eliminated, to transform all heat into work. The maximum possible percentage which may be transformed is given by Equation A-17.

Heat is transformed into mechanical or electrical energy by use of a heat engine. A steam turbine power plant is a heat engine as are, in the broadest sense, gasoline engines, gas turbines, and diesel engines. So are such direct energy conversion devices as thermoelectric converters, thermonic converters and magnetohydrodynamic (MHD) generators. These latter devices will now be discussed briefly. If one end of a material, such as a semi-conductor, is heated a voltage differential is set up between the two ends, thus offering the potential of delivering power to an external source. This device is known as a *thermoelectric converter*. When metals are heated to very high temperatures, electrons are emitted from the free surface. These electrons may be captured and diverted to an external load, thus constituting a *thermonic converter*. When a highly ionized gas (at very high temperature) is passed through a very strong magnetic field, electrons may be separated out and divered to an external load, thus constituting a *magnetohydrodynamic generator*. In a *fuel cell*, chemical energy is transformed directly into electrical energy. In *solar cells*, the energy of the photons in the solar beams is transformed directly into electrical energy. As such, these two devices are not heat engines and hence their efficiencies are not limited by the Carnot concept as given in Equation A-17.

The elements involved with a heat engine are shown in Figure A-6. The elements are (1) the heat source which supplies the heat, (2) the engine itself, and (3) the heat sink to which heat is rejected. In a fossil fuel steam power plant, the heat source is the hot gas resulting from combustion, the steam turbine is the heat engine and the condenser is the heat sink. In a nuclear steam power plant, the heat source is the nuclear reactor (see Figure A-7).

In the large fossil fuel steam power plant, steam is generated in the steam generator at high pressures (normally 2,400 to 3,500 psia) and is heated to

---

* This statement does not include the large amount of solar energy which is used by vegetation.

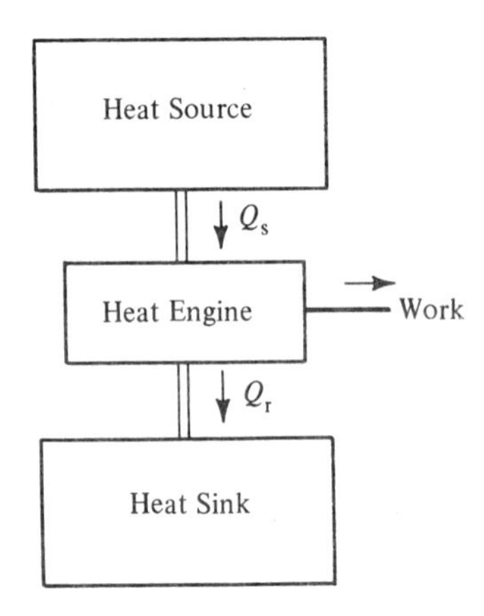

Figure A-6. Heat engine

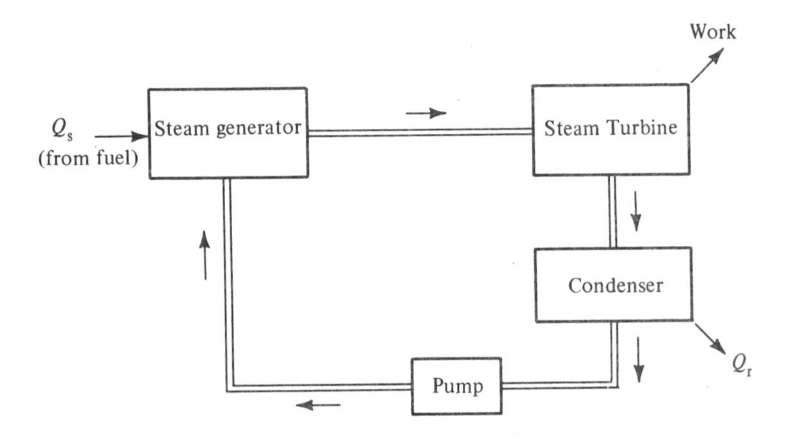

Figure A-7. Elements of a steam power plant

about 1,000°F.* The steam expands in the turbine to a very low pressure, frequently below 1 psia. From the turbine, the steam passes into the condenser where it gives up vast quantities of heat as it condenses. The condensation is caused by the water which is circulated through the condenser to absorb the heat given up by the condensing steam. This water frequently is called the circulating water. Of course the temperature of the circulating water leaving the condenser must be lower than that of the condensing steam, i.e., generally is below 95°F except in summer in the warmer regions. Although the circulating water leaving the condenser contains vast quantities of energy, because of its

* It should be noted that the boiling and condensing temperatures of steam depend on the pressure. At standard atmospheric pressure of 14.696 psia, water boils and condenses at 212°F. At 1 psia, the boiling temperature is 101.7°F. At a pressure of 2,800 psia, water boils at a temperature of 685.16°F.

low temperature it is not feasible to use a significant portion of this energy except in very special circumstances. It is true that if the steam were not expanded to as low a pressure in the turbine, its temperature in the condenser would be higher and the circulating water temperature at condenser exit would be sufficiently high that it could be used for various heating purposes. However, the work done in the turbine per pound of steam would be reduced significantly. Much more fuel would have to be burned to produce a given amount of electrical energy.

Most of the nuclear reactors used in this country produce steam at pressures between 1,000 and 1,500 psia. This steam is not heated above its boiling temperature. Because of these lower steam pressures and temperatures, our present nuclear plants have a significantly lower efficiency than the fossil fuel steam power plant.

In a strict sense, gas turbines and internal combustion engines are not heat engines since they use the hot combustion gases directly rather than to use these gases to supply heat to a heat engine. However, the net result is the same as if the gases, used in the engines, were to be heated by an external source, rather than by internal combustion. Hence the Carnot limitation on the thermal efficiency may be applied here.

A.6 REFRIGERATION

Since much energy is used for refrigeration and air conditioning, it is desirable to examine, briefly, some of the principles. Most of our refrigeration is obtained by use of the vapor compression system. In the broad sense, this system is a reversed heat engine (see Figure A-8). Assume that the engine is operated as a heat engine. Let 100 Btu be supplied from the source and 20 Btu of work produced. Then 80 Btu will be rejected to the heat sink. Now reverse

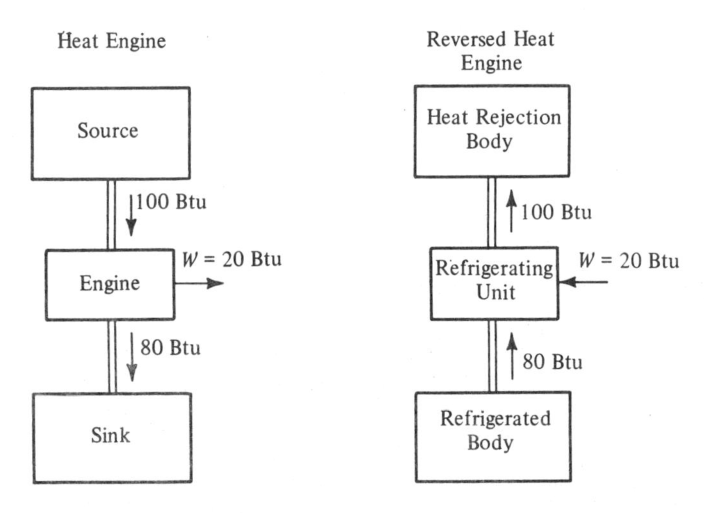

Figure A-8. Reversed heat engine for refrigeration (theoretical)

the operation of the engine. Put the 20 Btu of work back into it. Theoretically the 80 Btu of heat will be removed from the heat sink and 100 Btu will be delivered back into the heat source. Actually, because of losses, less than 80 Btu will be removed from the sink and less than 100 Btu will be delivered to the source. Because heat is removed from the sink, it will be kept cool, i.e., it will be refrigerated.

The elements of a vapor compression refrigeration system are shown in Figure A-9. The compressor takes vapor from the evaporator, compresses it to high pressure and delivers it to the condenser. Because the vapor is under high pressure in the condenser, it will condense as either water or air sweeps over the condenser to remove heat. The liquid formed in the condenser flows through the expansion or control valve, which greatly reduces the pressure at which it enters the evaporator. Because of the low pressure in the evaporator, the refrigerant evaporates as it picks up heat, thus producing the refrigerating effect.

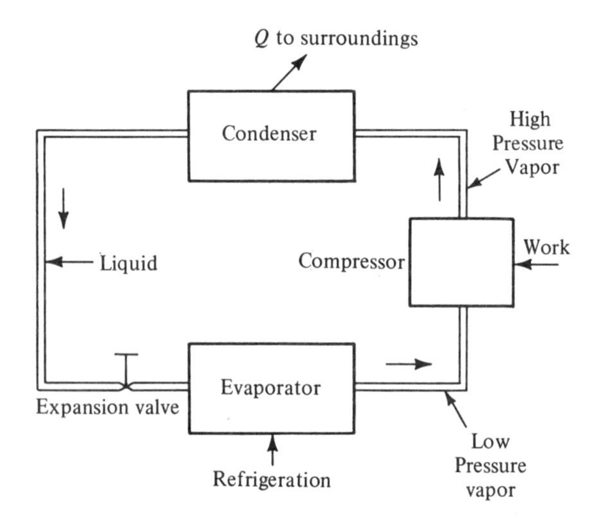

Figure A-9. Elements of a vapor compression refrigerating system

The common air conditioning unit is a refrigerating system. The liquid refrigerant, as it vaporizes in the evaporator, picks up heat from the air in the building which is being air conditioned. The compressor compresses the vapor and delivers it to the condenser. For smaller units, the condenser is located outdoors. Air is blown over the condenser, thus condensing the vapor. The liquid thus formed is returned to the evaporator.

A.7 HEAT PUMPS

In a refrigerating system, Figure A-8b and Figure A-9, heat is pumped from a low temperature body and is delivered, along with the work which has been put into the system for pumping purposes, to a higher temperature body. Thus the refrigerating system is, at the same time, a heat pump. The only fundamental difference between a refrigerating system and a heat pump is the purpose of the system—is it to refrigerate or is it to heat?

Since the heat pump normally can deliver two to three times as much heat as the heat equivalent of its electrical input, there is a growing use of heat pumps. Other reasons for this growth are the large increase in the cost of heating oil and also the fear that there will be a large increase in the cost of natural gas.

A heat pump has two distinct disadvantages in very cold weather. For this condition, the difference between the temperature of that in the evaporator and in condenser is large. Hence the heat pump requires a large electrical input to pump heat over this large temperature range.

With a low outside temperature, the temperature and hence the vapor pressure in the evaporator is low. With a low evaporator pressure the specific volume of the vapor leaving the evaporator and entering the compressor is large. Normally the compressor can handle a given volume of refrigerating vapor in a given time. Since the total volume per unit time equals the product of specific volume and the mass rate of flow, when the specific volume is high, the mass rate of flow is low. With a smaller mass rate of flow, much less heat can be pumped. Thus, unless the heat pump is made very large, it cannot deliver the amount of heat required in very cold weather. It is common practice to augment the heat pump output with direct electrical heating in very cold weather. Of course, this is expensive. A considerable amount of effort is being devoted to overcoming these two drawbacks of heat pumps.

In those regions of the country where very cold weather is seldom encountered, the overall cost of heating with a heat pump definitely is less than that with oil. If, as anticipated, there is a large increase in the price of natural gas, then heat pumps can become competitive with natural gas in many regions of the country.

It should be noted that, by reversing some valves, a heat pump can be operated as an air conditioning unit to provide summer cooling.

SUPPLEMENT A

ILLUSTRATIVE EXAMPLES

Example A-1. Determine the force, lb_f, required to accelerate 12 lb_m at the rate of 20 ft/sec^2.

Solution. From Equation A-1a,

$$F' = \frac{ma}{g_c} \text{ or}$$

$$F' = \frac{12}{32.174} \times 20 = 7.46 \text{ lb}_f \qquad \text{answer}$$

Example A-2. A pressure gauge indicates a pressure of 280 psi. The barometric pressure is 29.42 in. of mercury. Determine the absolute pressure.

Solution. At normal room temperatures, one inch of mercury is equivalent to 0.491 psi. Then the absolute pressure equals:

$$280 + 29.42 \times 0.491 = 294.4 \text{ psia} \qquad \text{answer}$$

Example A-3. Convert a temperature of 80°F to degrees K.

Solution. From Equation A-3b,

$$°\text{C} = (80 - 32)\frac{100}{180} = 26.67°\text{C}$$

From Equation 2-4b,

$$°\text{K} = 26.67 + 273.15 = 299.82°\text{K} \qquad \text{answer}$$

Example A-4. How much heat, in Btu, is required to heat a tank of 60 lb of air from 70°F to 180°F?

Solution. From Equation A-6,

$$Q = mc_v(T_2 - T_1)$$

Since the volume of the tank remains constant, the constant volume specific heat c_v is to be used. Its value for air is 0.1715 Btu/lb-°F. Hence,

$$Q = 60 \times 0.1715(180 - 70) = 1{,}132 \text{ Btu} \qquad \text{answer}$$

Example A-5. Air at a constant pressure of 74.5 psi moves a 6 in. diameter piston a distance of 2.2 ft. Determine the work done.

Solution. The area of the piston equals

$$\frac{6^2 \times \pi}{4} = 28.27 \text{ sq in.}$$

Using Equation 2-7,

$$\text{Work} = 74.5 \times 28.27 \times 2.2 = 4{,}664 \text{ ft lb} \qquad \text{answer}$$

Example A-6. Determine the kinetic energy of a one and a half ton automobile travelling at 55 miles per hour.

Solution. The mass of the car equals

$$2{,}000 \times 1.5 = 3{,}000 \text{ lb}$$

The velocity of the car equals

$$\frac{55\times5{,}280}{3{,}600}=86.67\text{ ft/sec}$$

Using Equation A-9,

$$\text{K.E.}=\frac{3{,}000\times80.67^2}{2\times32.174}=303{,}400\text{ ft lb}\qquad\text{answer}$$

Example A-7. A power plant burns 255 tones of coal per hour when delivering 760,000 kW. The heating value of the coal is 13,200 Btu per lb. Determine the thermal efficiency of the plant.

Solution. From Equation A-16,

$$\eta_t=\frac{760{,}000\times3{,}413\times100}{13{,}200\times255\times2{,}000}=40.9\text{ percent}\qquad\text{answer}$$

where 3,413 is the Btu per kWh.

Note: This is a reasonable value of efficiency for this type of plant.

Example A-8. The mean temperature at which heat supplied to a heat engine is 400°F. Heat is rejected at 120°F. Determine the maximum possible thermal efficiency.

Solution. The maximum possible thermal efficiency is that of the Carnot cycle. Hence, from Equation A-17,

$$\eta_t=\frac{400-120}{400+460}=32.6\text{ percent}\qquad\text{answer}$$

Note: The conversion factor from °F to °R has been rounded off to 460 since it is difficult to determine temperatures closer than 1°F except in the laboratory.

Example A-9. Determine the minimum amount of power required for a heat pump to deliver 50,000 Btu per hour at a temperature of 120°F when receiving heat at 50°F.

Solution. The minimum amount of power required will be more than for a Carnot cycle system. For a Carnot cycle heat engine, for the given temperatures, the work delivered when 50,000 Btu is added to it is, by Equation A-17,

$$50{,}000\,\frac{120-50}{120+460}=6{,}035\text{ Btu per hour.}$$

Hence when the system is reversed, a work input of 6,035 Btu will deliver 50,000 Btu to the high temperature body.

$$\frac{6{,}035}{3{,}413} = 1.768 \text{ kW} \qquad \text{answer}$$

where 3,413 is the Btu per kWh.

Note: The actual heat pump, because it cannot work on the Carnot cycle, may require an input of 4 to 5 kW to deliver 50,000 Btu per hour for the given conditions.

SUPPLEMENT B

PROBLEMS

A-1. A force of 40 lb_f is applied to a mass of 20 lb_m. Determine the acceleration in feet per second.

A-2. Determine the gauge pressure, psig, when the absolute pressure is 170.2 psia and the barometric pressure is 29.67 in. of mercury.

A-3. Convert −40°C to °F.

A-4. 2,700 Btu are added to a tank containing 130 lb of water at 80°F. Determine the final temperature of the water. Take the specific heat of water as 1.0 Btu/lb-°F.

A-5. a. An 8 in. diameter cylinder contains water under a pressure of 110 psi. Determine the volume of water which can be pushed out of the cylinder when 440 ft lb of work are used for this purpose. Assume that the pressure remains constant.
b. Is it necessary to know the diameter in part a?

A-6. A 2 ton car, starting at rest, climbs a 250 ft hill. At the top of the hill, the car has a velocity of 50 miles per hour. Determine the total energy possessed by the car at the top of the hill in excess of that at the bottom.

A-7. A 6 ton elevator rises 98 ft in 2 minutes. Determine the horsepower required to raise the elevator.

A-8. A steam power plant burns 750 barrels of oil per hour. The efficiency of the plant is 40.5 percent. The heating value of the oil is 140,200 Btu per gallon. There are 42 gallons per barrel. Determine the output of the plant in kW.

A-9. The mean tempeature at which heat is supplied to a heat engine is 400°F. Heat is rejected at 80°F. Determine the maximum possible thermal efficiency. Compare your answer with that of Example A-8.

A-10. A Carnot cycle engine produces 100 Btu of work when receiving heat at 540°F and rejecting it at 40°F. Determine the amount of work which can be produced by a second Carnot engine receiving the same amount of heat at 140°F and rejecting it at 40°F.

A-11. Same as Example A-9 except the heat pump receives heat at 10°F. Compare your answer with that of Example A-9.

A-12. Heat is added to the fluid in a cylinder of an engine. Is the change in internal energy of the fluid equal to the heat which has been added? Why?

A-13. Determine the number of kilowatts required to produce the heating of 50,000 Btu per hour in Example A-9 if no heat pump is used but the heating is provided by electrical heating elements.

A-14. Steam enters a turbine with an enthalpy of 1,450 Btu per pound and leaves with an enthalpy of 910 Btu per pound. The change in enthalpy equals the work delivered to the shaft of the turbine. Ninety-seven percent of this work is delivered as electrical energy by the generator attached to the turbine. Determine the steam flow per hour if the generator delivers 750,000 kW.

Appendix B

The International System (SI) of Units

To bring order out of the chaos which existed because of the large number of inconsistent units of measurements, an International Metric Commission met in Paris on May 20, 1875. This meeting led ultimately to the establishment of the meter-kilogram-second (MKS) system with the metric system being used for the units. The metric system is a decimal system in which the various units used for measuring a given quantity are multiples of ten of each other. In 1960, the Eleventh General Conference of Weights and Measures, recognizing that some deficiencies had developed in the MKS system, established the International System of Units (SI). Thirty-six countries, including the United States, participated in this conference. Basically the SI system defines six fundamental units and two supplementary units:

Unit	Name of Unit	Symbol
Length	meter	m
Mass	kilogram	kg
Time	second	s
Electric current	ampere	A
Temperature	kelvin	K
Luminous intensity	candela	cd

Supplementary Units:

Plane angle	radian	rad
Solid angle	steradian	sr

To avoid the use of very large or very small numbers, prefixes have been established to indicate orders of magnitude:

Prefix	Symbol	Multiplication Factor
terra	T	10^{12}
giga	G	10^{9}
mega	M	10^{6}

kilo	k	10^3
hecto	h	10^2
deka	da	10^1
deci	d	10^{-1}
centi	c	10^{-2}
milli	m	10^{-3}
micro	μ	10^{-6}
nano	n	10^{-9}
pico	p	10^{-12}
femto	f	10^{-15}
atto	a	10^{-18}

For example, a kilogram is 10^3 times a gram or one thousand grams. A megawatt is 10^6 times a watt or a million watts. Since a kilowatt is 10^3 times a watt, the megawatt is a thousand kilowatts.

Four named derived units are

Unit	Name of Unit	Symbol
Force	Newton	$N=1$ kg m/s^2
Work, energy, heat	joule	$J=1$ N m
Power	watt	$W=1$ J/s
Electrical potential	volt	$V=1$ W/A

All other units, unnamed, are expressed in terms of the named unit. Some of these are

Unit	SI Unit	Symbol
Area	square meter	m^2
Volume	cubic meter	m^3
Density	kilogram per cubic meter	kg/m^3
Pressure	newtons per square meter	N/m^2
Velocity	meters per second	m/s
Specific heat	joules per degree kelvin per kilogram	J/K-kg

Compared with the English system of units, the SI system is very simple. The English system involves a very large number of conversion constants, such as 16 ounces per pound, 43,560 square feet per acre, 5,280 feet per mile and 778.16 foot pounds per Btu. In contrast, conversion in the SI system simply involves multiples of ten. The simplicity of the SI explains why substantially all countries of the world, with the exception of the United States, are now using the SI system. We are in the process of adopting the SI system; however, the transition is slow. It is difficult for us to think in terms of grams, meters, Celsius temperatures, joules, and newtons. For this reason the units used in this text are the English units. But the transition to the SI system definitely will occur. This discussion of the SI system is presented here in order for the reader to understand and use the conversion factors which follow.

Appendix C

Conversion Factors

LENGTH

12 inches = 1 foot
3 feet = 1 yard
5,280 feet = 1 mile
6,076 feet = 1 nautical mile

2.54 centimeters = 1 inch
30.48 centimeters = 1 foot
3.28084 feet = 1 meter
1,609.3 meters = 1 mile
10^8 angstroms = 1 centimeter
1 hectometer = 100 meters

AREA

144 square inches = 1 square foot
9 square feet = 1 square yard
43,560 square feet = 1 acre
640 acres = 1 square mile

6.4516 square centimeters = 1 square inch
4,046.856 square meters = 1 acre
1 square hectometer = 1 hectare
1 hectare = 10,000 square meters

VOLUME

1,728 cubic inches = 1 cubic foot
231 cubic inches = 1 U.S. gallon
42 gallons = 1 barrel (petroleum)
7.4805 gallons = 1 cubic foot

1,000.028 cubic centimeters=1 liter
28.316 liters=1 cubic foot
3.7853 liters=1 U.S. gallon

MASS

16 ounces=1 pound (avoirdupois)
2,000 pounds (avoirdupois)=1 ton
7,000 grains=1 pound (avoirdupois)

453.5924 grams=1 pound (avoirdupois)
2,204.622 pounds=1 metric ton=1,000 kilograms
2.204622 pounds=1 kilogram
32.1740 pound mass=1 slug

PRESSURE AND FORCE

14.6960 pounds per square inch=1 standard atmosphere
29.921 inches mercury at 32°F=1 standard atmosphere
0.43309 pounds per square inch=1 foot water at 60°F
0.4898 pounds per square inch=1 inch mercury at 0°F
32.1740 poundals=1 pound force

$$1 \text{ newton (N)} = \frac{1 \text{ kilogram meter}}{\text{seconds}^2}$$

$$1 \text{ pascal (Pa)} = \frac{1 \text{ N}}{\text{meter}^2}$$

1 pound/square inch=6,894 Pa
1 bar=10^5 Pa=14.505 psi

TEMPERATURE

°F=°C×1.8+32
°C=(°F−32)×1.8
°K=°C+273.15
°R=°F+459.67

ENERGY AND POWER

778.16 foot pounds=1 Btu
33,000 foot pounds per minute=1 horsepower
2,544 Btu per hour=1 horsepower
3,413 Btu per hour=1 kilowatt
252 calories=1 Btu
860 calories=1 watt hour
4.1868 joules=1 calorie
10^7 ergs=1 joule
745.7 watts=1 horsepower
1 joule=1 watt second

Appendix D

Properties

BOILING WATER

Pressure, psia	Boiling Temperature, °F	Density lb/cu ft	Specific heat Btu/lb-°F (C_P)
0.0886	32	62.41	1.005
0.1170	39	62.42	1.004
0.2563	60	62.36	1.001
0.9503	100	62.00	0.997
3.722	150	61.19	0.999
11.529	200	60.09	1.005
14.696	212	59.83	1.008
66.98	300	57.31	1.03
247.1	400	53.65	1.08
680.0	500	48.95	1.18
1,541.0	600	42.32	1.52
3,090	700	27.28	
3,204	705.4	19.79	

Note: The pressure at any given temperature below 500°F may be increased several hundred pounds without significantly changing the density or specific heat. For temperatures above 500°F, refer to "Steam Tables," Keenan, Keyes, Hill and Moore, John Wiley and Sons. The above data was based largely on these tables.

AIR

The density and specific heat of air at the standard pressure of 14.696 psia is

Temperature, °F	Density lb/cu ft	Specific heat Btu/lb-°F (C_P)
0	0.0862	0.240
77 (25°C)	0.0739	0.240
100	0.0708	0.240
200	0.0601	0.241
300	0.0522	0.243
500	0.0413	0.248
750	0.0328	0.255
1,000	0.0272	0.263
2,000	0.0161	0.276

The density of air at other pressures below 1,000 psia may be calculated by multiplying the density given here, at the desired temperature by $0.068 \times p$ where p is the pressure at which the density is desired, psia. For example, what is the density of air at 300°F, 400 psia? From the above table, the density at 300°F is 0.0522 lb/cu ft. The desired density$=0.0522 \times 0.068 \times 400 = 1.420$ lb/cu ft.

OTHER GASES

To determine the density of a gas at any given pressure and temperature, determine the density of air for the given conditions and multiply by the following factors:

Carbon monoxide	0.967
Hydrogen	0.0696
Oxygen	1.105
Nitrogen	0.967
Helium	0.276

Note: Tables must be used to determine the density of vapors such as steam and various refrigerants except at very low pressures.

Symbols

a	acceleration
c	specific heat
c_p	specific heat at constant pressure
c_v	specific heat at constant volume
d	distance
D	nucleus of a deuterium atom
f or F	force
g	acceleration of gravity
g_c	gravity dimensional constant $\left(=32.174 \frac{\text{lb}_m\text{ft}}{\text{lb}_f\text{sec}^2}\right)$
H	enthalpy
h	heat transfer coefficient
K.E.	kinetic energy
m	mass
η	efficiency
η_t	thermal efficiency
n	number of years (for sinking fund factor)
P.E.	potential energy
r	interest rate, percent
Q	heat
T	temperature
T	nucleus of a tritium atom
U	internal energy
V	velocity
V	volume
W	weight
W	work

Abbreviations

AEC	Atomic Energy Commission
bbl	barrel
Btu	British thermal unit
cfs	cubic feet per second
cm	centimeter
°C	degrees Celsius
°F	degrees Fahrenheit
°K	degrees Kelvin
°R	degrees Rankine
d	distance
DOE	Department of Energy
EPA	Environmental Protection Agency
ERDA	Energy Research and Development Administration
FEA	Federal Energy Administration
ft	feet
HTGR	high temperature gas reactor
hp	horsepower
kW	kilowatts
kW_e	kilowatts electrical
kW_t	kilowatts thermal
kWh	kilowatt hours
lb	pounds
lb_f	pounds force
lb_m	pounds mass
LNG	liquified natural gas
m	meters
MHD	magnetohydrodynamics
MW	megawatts
MW_e	megawatts electrical

MW_t	megawatts thermal
MWh	megawatt hours
NASA	National Aeronautics and Space Administration
OTEC	ocean thermal energy conversion
ppm	parts per million
psi	pounds force per square inch
psia	pounds force per square inch absolute
psig	pounds force per square inch gauge
Quad	a quadrillion Btu (10^{15})
rpm	revolutions per minute
sec	seconds
TVA	Tennessee Valley Authority

Index

A
Acid rain, 267
Air preheaters, 31
Alternative energy sources, 5, 101
Atomic mass unit, 73
Atomic number, 73

B
Barrels of oil equivalent, 17
Biomass resources, general, 179
Biomass resources, specific
  Algae, 184
  Kelp, 185
  Manure, 184
  Plankton, 185
  Sewage, 184
British thermal unit (Btu), 11, 298

C
Calorie, 298
Carnot cycle, 303
  Carnot cycle thermal efficiency, 303
Catalyst, 204
Chemical energy, 300
Coal, 47
  Anthracite, 47
  Bituminous, 47
  Lignite, 47
  Subbituminous, 47
Coal, burning of, 52
Coal gasification, 201
  Future of, 209
  Present status, 208
  Problems involved, 208
  Underground (in situ) gasification, 205
Coal, its nature, 47
Coal liquefaction, 206
  Present status, 207
Coal mining, 51
  Strip mining, 51
  Underground mining, 51
Coal resources, 49
Cogeneration, 226
Condensate, 27
Condenser, elementary steam, 30
Conservation of energy, 291
Conservation of energy in electric power generation, 242
Conservation of energy in industry, 243
Conservation of energy in passenger transportation, 247
Conservation of energy in residential and commercial usages, 249
Conservation of energy in transportation, 245
Crisis. Are we approaching one?, 274
Cycles, 303

D
Deterioration of life-styles, 8
Deuterium, 190
Deutron, 190
Development of additional energy supplies, 9
Dilemmas in energy, 275
Dual cycles for power production, 224

E
Economizers, 31
Einstein energy equation, 71
Electrical energy, 299
  Importance of, 25
  Production, 26
Electrical generating capacity, 26
Electron volt, 73
End use of electrical energy, 21
Energy consumption by electrical utilities, 20
Energy, definition of, 11, 293
Energy demands, 18
Energy, forms of, 297
Energy from the oceans, 147
Energy options, 6
Energy reserves, 45
  Possible reserves, 45
  Probable reserves, 45
  Proved reserves, 45
Energy resources distribution, 18
Energy situation at present, 5
  Energy situation, an overview, 273
Energy use distribution, 19
Energy use in transportation, 246
Enthalpy, 301
Environmental problems with coal production, 260
Environmental problems with energy transformations, 263
Environmental problems with energy utilization, 266
Environmental problems with geothermal energy, 261
Environmental problems with hydropower, 261
Environmental problems with utilization of energy of the oceans, 263
Environmental problems with oil and natural gas production, 260
Environmental problems with shale oil production, 261
Environmental problems with solar energy utilization, 262
Environmental problems with uranium production, 261
Environmental problems with wind power, 262
Environments vs energy demands, 259

F
Factors causing the energy problems, 5
Fast breeder reactors, 80
Feedwater, 28
Feedwater heaters, 30
Financing energy supplies, 9
First law of thermodynamics, 11, 302
Fissile, 72
Fission, 72
Fixed costs, 36
Force, definition, 294
Fuel cells, 230
Fuel cell developments, 232
Fuel cell efficiencies, 232
Fuel cell reactions, 231
Furnaces of steam generators, 31
Fusion, (nuclear), 189
Fusion power plant, 194
Fusion process, 189
Fusion reactions, 191
Fusion reactor safety, 195

G
Gas producers, 201
Gasification concept, 202
  Coal, 201
  Coal, underground (in situ), 205
  Oil, 206
Gasohol, 183
Geothermal energy, 133
Geothermal hot rocks energy, 140
Geothermal hot water power, 136
  Environmental problems, 138
Geothermal steam power, 135
  Environmental problems, 136
Gravity constant, g_c, 294

H
Half life, radioactive substances, 79
Head, for hydropower production, 89
Heat, definition of, 298
Heat engines, 305
Heat pumps, 308
Heating values, 12
  Higher or gross, 12
  Various types of coal, 48
Hydraulic turbine horsepower, 90
Hydraulic turbine types, 89
Hydrogen economy, 215
Hydrogen for energy storage, 215
Hydrogen energy utilization, 217
Hydrogen production, 214
  Electrolysis of water, 214
  Thermal-chemical method, 214
Hydrogen for transportation power, 216
Hydropower, 89

Hydropower, present and future
  capacities, 91
Hydroelectric power, pros and cons, 92
Hydropower production, 89

I
Incineration of refuse, 172
  For power production, 173
  For steam production, 174
Increase energy demand factors, 16
Industrial (intermediate) gas from
  coal, 204
Initial costs of power plants, 38
Internal energy, 300
Isotopes, 72

K
Kinetic energy, 300

L
Lasers (for nuclear fusion), 195
Liquified natural gas (LNG), 17

M
Mass, definition, 293
Magnetic containment (for nuclear
  fusion process), 192
Magnetohydrodynamics (MDH) power
  production, 227, 305
Magnetohydrodynamics design and
  operational problems, 229
Municipal refuse, 171
Municipal refuse composition, 171

N
Natural gas, 65
Natural gas composition, 65
Natural gas resources, 67
Natural gas use, 66
New conversion methods, 223
Newton, 293
Nuclear energy, 71, 300
Nuclear fission, 71
Nuclear fission reactors, 76
  Boiling water, 77
  Control rods, 74
  Heavy water, 76
  High temperature gas, 76
  Light water, 76
  Moderators, 73
  Pressurized, 76
Nuclear fuel preparation, 74
Nuclear fusion process, 72
Nuclear fuel reprocessing, 78
Nuclear fusion, 72
Nuclear reactor safety, 83
Nuclear waste disposal, 79
Nucleus of atom, 72
  Protons, 72
  Neutrons, 72

O
Ocean energy potentials
  Ocean currents, 149
  Salinity gradients, 153
  Thermal gradients, 150
  Tides, 148
  Waves, 148
Oil, amount recoverable, 62
Oil, composition, 59
Oil, formation of, 59
Oil gasification, 206
Oil, importance, 60
Oil, reserves, 61
Operating costs, 39
Options for the solution of the
  energy problems, 6

P
Particulate matter (ash), 52
Peat, 54
Pipeline (high Btu) gas from coal, 204
Plasma, 190
Potential energy, 300
Power, definition, 301
Power gas composition, 202
Power (low Btu) gas from coal, 202
Power gas uses, 203, 204
Pressure, 295
  Absolute pressure, 295
  Barometric pressure, 13
  Gage pressure, 295
Proximate analysis, 47
  Ash, 47
  Fixed carbon, 47
  Moisture, 47
  Volatile, 47
Pulverized coal, 31
Pumped storage, 93
Pyrolysis of coal, 206
Pyrolysis of refuse, 175
  Efficiency, 175
For gas production, 175
For oil production, 175

Q
Quad, 17

R
Rankin cycle, 27
Refrigeration, 307
  Vapor compression system, 308
Reversible process, 304

S
Second law of thermodynamics, 302
Shale oil, 159
Shale oil production, 160
Shale oil production problems, 163
Shale oil resources, 160
Sinking fund factor, 37
Specific heat, 298
Solar (photovoltaic) cells, 117
Solar cell efficiencies, 117
Solar-electrical energy storage, 119
Solar cell space power, 118
Solar collectors
  Flat plate, 106
Solar energy, 103
Solar energy, air conditioning, 114
Solar energy, daily insolation, 104
Solar energy, direct heating and cooling, 105
Solar energy dissipation, 103
Solar energy land requirements, 105
Solar energy resource, 103
Solar energy storage, 111
Solar furnaces, 105
Solar heating of buildings, 111
  Costs, 113
Solar heating, passive, 116
Solar hot water heating, 106
  Costs, 110
Solar pond power production, 120
Solar "power tower" system, 120
Solar-steam power systems, 119
Solvent refining of coal, 207
Steam condensers, 30
Steam generators, 27
Steam power plants, 27
Steam turbines, 28
Steam turbine work as affected by exhaust pressure, 29
Sulfur in coal, 48
Sulfur dioxide controls, 53
Systems, concept of, 298

T
Tar sands, 165
Tar sands oil production, 166
Tar sands resources, 165
Tar sands, underground (in situ), 167
  Oil production, 167
Temperature, 12
  Absolute temperatures, 12
    Kelvin, 13
    Rankine, 13
  Celsius temperature, 13
  Fahrenheit temperature, 12
Thermal efficiency, 302
Thermal ocean gradient power, 150
Thermal ocean gradients power plants, 151
Thermal ocean gradients power plant materials, 152
Thermal ocean gradients power utilization, 152
Thermal pollution, 261
Thermoelectric energy conversion, 305
Thermionic energy conversion, 305
Thorium and the slow breeder reactor, 81
Three Mile Island, 84
Tidal power, 148
Tritium, 191

U
Ultimate analysis of coal, 48
Uranium, 73
  Composition, 73
Uranium resources, 82

V
Vegetation for energy production, 182
Vegetation wastes for energy production, 182

W
Water wheels for power production, 89
Weight, 295
Wind power, 125
Wind power energy utilization, 129
Wind power future, 127
Wind power—larger scale, 126
Wind power offshore, 129
Wind power potential in the Great Plains, 128
Wind-wave power, 148
Wood for energy production, 179
Wood potential for energy production, 180
Wood wastes for energy production, 180
Work, definition, 299

About the Author

Born in Bethany, Connecticut in 1903, Jesse S. Doolittle has spent the past 55 years consulting, writing and teaching. Presently a Professor Emeritus and consultant to the Mechanical and Aerospace Engineering Deparment of North Carolina University, Professor Doolittle started his teaching career at the Case Institute of technology. He then joined Penn State University for 16 years before becoming affiliated with North Carolina State University. During his career, Professor Doolittle has written many papers and nine text books, including his latest title, *Energy: A Crisis, A Dilemma Or Just Another Problem?, 2nd Edition*. He has acted as a consultant to the U.S. Navy, the U.S. Department of Interior, and the U.S. Public Health Department, and is currently a consultant to Carolina Power and Light.

Jesse S. Doolittle has been recognized as one of the "Outstanding Educators in America" along with "American Men of Science", "Engineers of Distinction" and "Who's Who in Engineering."